Essential Algebra for Chemistry Students

Second Edition

David W. Ball

Professor of Chemistry
Cleveland State University
Cleveland, Ohio

BROOKS/COLE
CENGAGE Learning

Australia • Brazil • Japan • Korea • Mexico • Singapore • Spain • United Kingdom • United States

BROOKS/COLE
CENGAGE Learning™

**Essential Algebra for Chemistry Students,
Second Edition**
David W. Ball

For product information and technology assistance,
contact us at
**Cengage Learning Customer & Sales Support,
1-800-354-9706.**

For permission to use material from this text or product,
submit all requests online at
www.cengage.com/permissions.
Further permissions questions can be emailed to
permissionrequest@cengage.com.

Library of Congress Control Number: 2004115550

ISBN-13: 978-0-495-01327-3
ISBN-10: 0-495-01327-7

Brooks/Cole
20 Davis Drive
Belmont, CA 94002-3098
USA

Cengage Learning is a leading provider of customized learning solutions with office locations around the globe, including Singapore, the United Kingdom, Australia, Mexico, Brazil, and Japan. Locate your local office at: **www.cengage.com/global**.

Cengage Learning products are represented in Canada by Nelson Education, Ltd.

To learn more about Brooks/Cole, visit
www.cengage.com/brookscole.

Purchase any of our products at your local college store or at our preferred online store **www.cengagebrain.com**.

Printed in the United States of America
11 12 13   15 14 13

No writing project is the effort of just one person. Several other people have been instrumental in the development of this project, and they all have my thanks. For the first edition, Keith Dodson of West Educational Publishing Company provided timely feedback and guidance. Rick Mixter, also from West, and Jennifer Welchans and Trent Carruthers, the local West representatives, were very supportive. Helen Newman was my first guinea pig and agreed to read the entire manuscript from cover to cover, worked the problems, and gave me an honest appraisal of the manuscript. Students Michelle Ballash, David Fuote, Sheri Salay, and Erine Stames class-tested the manuscript under real classroom conditions and made some valuable suggestions. For the second edition, many thanks to Regina Johnson and David Harris of Brooks/Cole Publishing for their support, and thanks to Gail R. Ball for technical assistance.

For Gail

Contents

Preface

To the Instructor

In my experience, many students come into their first chemistry sequence – either at the allied health or at the science/engineering level – with a relative inability to perform the basic mathematical skills that are necessary to do some of the material. Every year, I see students dropping general chemistry not because they don't understand the **chemistry**, but because they can't do the **math**. Some of them can't even use their own calculator correctly, even though the students think they can. Many of these students say that they have satisfied the math prerequisite for chemistry, which is usually high-school algebra. However, when asked to actually perform math (like on an exam), they can't.

This book is meant to serve as a review of essential algebra skills for the chemistry student. Although most students have had algebra, it may have been so long ago or the knowledge may not have been used since that time. Many students have deficient math skills and are not truly prepared for chemistry.

This book will help resharpen the math skills that are necessary for most introductory chemistry courses. It is not written to accompany any specific chemistry textbook, and can be considered a stand-alone book. It is an appropriate supplement for chemistry textbooks on the preparatory, allied health/general-organic-biological (GOB), and science/engineering general chemistry level. It does not cover trigonometry or calculus, since most general chemistry texts do not use these topics directly.

Each chapter has exercises worked out in detail and student exercises at the end. Answers to all of the student exercises are included with each chapter, but are intentionally on the left page so they aren't exposed unintentionally. Material that discusses calculator use has calculator key sequences in boldface and sans serif font, to distinguish it from the rest of the material.

The intended size of this book means that a lot of specific types of chemistry problems could not be covered directly. For example, percent yield calculations and limiting reagent problems are not discussed, nor are all of the huge variety and complexity of equilibrium constant problems. These problems can be found in a chemistry book. Hopefully, this book will help provide the confidence in the math skills that will be necessary to work these types of problems successfully.

New to the second edition are references to the University of Massachusetts' Online Web-Based Learning (OWL) material, which is being incorporated into all of Brooks/Cole's GOB and general chemistry texts. If you are using a Brooks/Cole text, you can get access to the OWL materials, which include some online math skills exercises. These exercises serve as additional practice of some of the skills in this book, and also provide the opportunity to learn to navigate the OWL system. References to OWL material, in bold and sans serif font, are embedded in the appropriate sections of this book.

To the User

Good math skills are essential in order to master the basics of any science, including chemistry. If you think that you are deficient in those skills, then this book should help you brush up on the necessary material you will need to learn chemistry.

You are encouraged to start at page 1 and go through **the entire book**. The later chapters build on material of the previous chapters, and there's a chance that some crucial point might be missed if a reader simply goes right to a later chapter. Also, a word to the wise: **work out the problems**. Don't just look at an exercise and say to yourself, "I know how to do that." Even worse: do not look at a problem, check the answer, and say to yourself, "I could have gotten that answer." If you do not actually work out the problem and see how to get to the correct answer, you are not learning anything. You learn best by doing. Do. Work out the problems yourself.

The book is (hopefully) straightforward, loaded with examples and exercises (and answers), and easy to read. There is space in the book for you to work out problems on the page, should you so choose. There are also some sections and problems that are set aside specifically for the use of calculators, because knowing how to use your calculator is an important part of understanding the math you use.

The ordering of the chapters mimics the order how some of the material is utilized during the course of most introductory chemistry texts. However, you may need to cover the first four or five chapters quickly, since concepts like balancing equations and the numerical problems that accompany balanced equations require a mastery of all of Chapters 1 – 5. The final chapter on graphing will be particularly useful to you if you are also taking a chemistry laboratory course, since it is common to perform an experiment that requires you to graph your results. It should also help you in your lecture courses to better understand graphs in your texts.

When I was in school, I had a math teacher say to me, as I recall, "I don't care if you don't like math; you'll be doing it the rest of your life." (Or something along those lines.) Well, she was right. And as useful as math is, it is even more important that it be done correctly.

Good luck, and good chemistry (and math) to you all!

Cleveland, Ohio
May 2004

Chapter 1. Numbers, Units, and Scientific Notation

Introduction

All of science is very mathematical, and math uses numbers. But science does not use numbers alone; it uses units to express quantities, too.

A <u>number</u> tells you <u>how</u> <u>much</u>; a <u>unit</u> tells you how much <u>of</u> <u>what</u>. In science, both numbers and units are very important. For example, when one is given the quantity "2 grams of iron", then the number is "2" and the unit is "grams of iron":

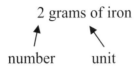

This may seem like a very trivial distinction, but later we will find that it is a very important one.

It is critical in chemistry to keep track not just of the numbers in a mathematical problem, but also the units. For example, when you work a mole-mole type of problem in chemistry, a frequent question you might have is "Moles of WHAT?" That's because the amount, moles, can refer to anything. Consider, for example, a dozen eggs, a dozen doughnuts, a dozen people, a dozen water molecules. If you were asked how many are present and you answer "A dozen" (meaning twelve), then your answer is incomplete. A dozen OF WHAT? Of course, the question is incomplete; you should be asked how many eggs you have, or how many people are present. In that case, the unit is understood. It's eggs, or people. But too often in chemistry problems, it's easy to lose track of what the exact unit is.

Example 1.1. Identify the number and unit in each of the given quantities. (a) A dozen eggs. (b) Three blind mice. (c) 60 miles per hour. (d) 8850 meters above sea level. (e) 365 days.

Solutions. (a) number = dozen (or 12), unit = eggs. (b) number = 3, unit = blind mice. (c) number = 60, unit = miles per hour. (d) number = 8850, unit = meters above sea level. (e) number = 365, unit = days.

More on Numbers

Numbers themselves can be greater than zero (<u>positive</u> numbers) or less than zero (<u>negative</u> numbers). Negative numbers have an explicit negative sign preceding them, like –4 or –200, whereas positive numbers may or may not have a positive sign. For example, the numbers 75 and +75 both refer to <u>positive</u> 75. **In the absence of any sign, always assume that the number is positive.** Zero itself is considered neither positive nor negative.

The numbers that we use to count things (1, 2, 3, 4, 5, ...) are called <u>whole</u> <u>numbers</u> or <u>integers</u>. Integers include 0 as well as negative numbers. The numbering system that we use allows us to combine single digits to make larger numbers to indicate how many 10's, how many 100's, how many 1000's, etc., a number represents. For example, the number 1234 implies one 1000's, two 100's, three 10's, and four 1's. The position of each digit indicates what it represents. We say that the 1 is in the thousands' place, the 2 is in the hundreds' place, the 3 is in the tens' place, and the 4 is in the ones' place:

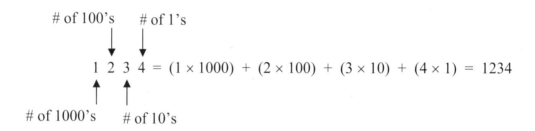

Objects that come in packages (like people, eggs, donuts, etc.) are easily counted using integers. However, there are many objects that can be divided into smaller parts. We use <u>decimal</u> <u>numbers</u> to indicate those smaller parts. We use a period, called a <u>decimal</u> <u>point</u>, to separate the integer numbers from the decimal numbers. (In Europe and elsewhere, they use a comma to separate integers from decimal numbers.) For example, if we have two grams plus one half of a gram of iron, we use the decimal number 2.5 to describe the number of grams of iron, where the ".5" is used to represent one half, or five tenths, of a gram. Position again determines what fraction a digit represents. If the position of a decimal number is next to the decimal point, it represents the number of $\frac{1}{10}$ths of a unit. If the position of a decimal number is the second position after the decimal, it represents the number of $\frac{1}{100}$ths of a unit, and so forth. Like above, then, we have for decimal numbers:

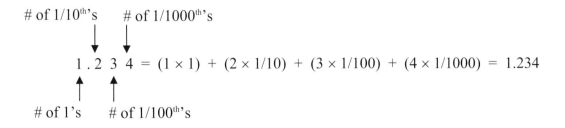

For either integer or decimal numbers, there may be a zero in any position to indicate that there is no contribution from that place. For example, the number 101.01 means "one hundred and one, and one hundredth." However, there are no tens and no tenths in this number. Because our numbering system uses the position of the number as an important way to communicate its value, we have to use zeros to properly position non-zero digits.

Example 1.2. Write the numbers from the following descriptions. (a) Three thousands, eight hundreds, 3 tens, and 9 ones. (b) Five ones, 7 tenths, and 2 hundredths. (c) Negative three ones, 7 hundredths, and 2 thousandths.

Solutions. (a) 3839 (b) 5.72 (c) –3.072. Notice in the third answer the "0" in the tenths' place to properly place the hundredths' and thousandths' places.

More on Units

As mentioned above, the unit attached to a number tells you what specific thing is being quantified. In chemistry, it is very important to keep track of the unit that goes along with the number.

Many people in the United States are very familiar with a particular set of units called English units: pounds, ounces, gallons, feet, miles, degrees Fahrenheit. While there is nothing wrong with these units, it has to be admitted that there is no consistency between them. Scientists all over the world have been using a different set of units. Most nations also have adopted them; the United States is one of the last holdovers using the English units.

Although the other set of units is sometimes called the metric system, the units are more properly called SI units (the SI comes from the French phrase "Le Système International d'Unités"). The SI units are divided into two groups: fundamental units and derived units. There are only seven fundamental units, five of which are commonly used in chemistry:

Fundamental Units

Quantity	Unit	Abbreviation
Mass	kilogram	kg
Amount	moles	mol
Time	second	s (but sec is common)
Temperature	Kelvin	K
Length	meter	m

(The other fundamental units are candela, for luminous intensity, and ampere, for electric current.) The fundamental units also have abbreviations, given in the table, in order to communicate them more efficiently. The kilogram is itself based on the unit gram, which has the abbreviation "g."

Derived units are combinations of the fundamental units. For example, to determine a volume of an object, you multiply its length times its width times its height. To calculate the volume, you not only multiply the numbers of the lengths you measure, but **you multiply the units as well**. If you express your units in meters, then the units of the volume become meter × meter × meter, or meter^3, or more simply m^3. The unit for volume is a derived unit.

Another example is the unit for velocity. Velocity is defined as "distance traveled divided by time taken." Therefore, in terms of fundamental units, velocity has the derived unit of $\dfrac{\text{meters}}{\text{second}}$, or $\dfrac{\text{m}}{\text{s}}$.

Some derived units are more complicated combinations (usually multiplications and divisions) of fundamental units, and it is common to define a new unit to stand for the more complicated combination. (You should never forget, however, that the "new" unit is actually a combination of the fundamental units.) The unit of force is called a newton (abbreviated N) but is a combination of kilogram, meter, and second units:

$$\text{newton} = \frac{\text{kilogram} \times \text{meter}}{\text{second}^2}$$

$$N = \frac{\text{kg} \cdot \text{m}}{\text{s}^2}$$

The following table lists other, common derived units and their "new" definitions. You will probably encounter most of them in your study of chemistry.

Derived Units

Quantity	Fundamental Units	"New" Unit	Abbreviation
Force	$\dfrac{kg \cdot m}{s^2}$	newton	N
Work, heat, energy	$\dfrac{kg \cdot m^2}{s^2} = N \cdot m$	joule	J
Power	$\dfrac{kg \cdot m^2}{s^3} = \dfrac{J}{s}$	watt	W
Temperature	$K - 273.15$	degree Celsius	°C

The last new unit, for temperature, is based on the fact that the absolute temperature scale, which uses the Kelvin temperature unit, is the same size as the Celsius (or Centigrade) degree, only shifted by a little over 273 degrees. Scientists commonly use the Celsius scale to express temperatures. At the same time, they recognize that many mathematical equations used in chemistry require that temperature be expressed in the Kelvin temperature scale. (But it's simple to change it.) Most nations (and again, the US is almost the only exception) use degrees Celsius to express temperatures as part of their weather report!

Although the SI unit for volume should be m³, a meter cubed (or "a cubic meter") is a rather large volume. The SI system therefore defines a more manageable unit, the liter, abbreviated L, which is one thousandth of a cubic meter, or 0.001 m³. Chemistry thus uses the liter (a volume that is a little larger than a US quart) as the "basic" unit of volume, even though it is ultimately considered a derived unit.

Having "manageable" units is a primary concern of the SI units. However, despite the definition of the liter, we can still easily imagine circumstances where this volume unit is too large or too small to effectively describe the volume. The same is true of the other fundamental and derived units. However, the SI system also has a series of prefixes that are used to indicate a scaling factor. The easy part of these prefixes is that they all involve either multiplication or division by some power of 10.

The common prefixes and their scaling factors are:

Scaling Prefixes for SI Units

Prefix	Scaling Factor	Abbreviation
giga-	$1{,}000{,}000{,}000 \times$	G
mega-	$1{,}000{,}000 \times$	M
kilo-	$1{,}000 \times$	k
deci-	$\frac{1}{10} \times$	d
centi-	$\frac{1}{100} \times$	c
milli-	$\frac{1}{1000} \times$	m
micro-	$\frac{1}{1{,}000{,}000} \times$	μ
nano-	$\frac{1}{1{,}000{,}000{,}000} \times$	n

There are others, but these are the most common ones that you will probably encounter in chemistry.

The prefixes are combined with the SI units to define "new" units that have more manageable numbers. For example, the fundamental unit of mass, the kilogram, has one of the prefixes: "kilo-". The unit "kilogram" means "1000 × grams", or 1000 grams:

$$k i l o g r a m$$

$$1000\times \quad \text{gram} \quad = \quad 1000 \text{ grams}$$

The prefixes also have abbreviations, and they are combined with the abbreviations of the units to create a simple abbreviation for the scaled unit. Therefore, the abbreviation for kilogram is "kg" (as indicated above), and it is simply the combination of the abbreviations for kilo- (k) and gram (g).

Example 1.3. State the meanings and give the abbreviations of the following units. (a) kilometer (b) microliter (c) megajoule (d) nanonewton (e) centimeter.

Solutions. (a) A kilometer is 1000 × meter or 1000 meters and has the abbreviation km. (b) A microliter is $\frac{1}{1{,}000{,}000}$th of a liter and has the abbreviation μL. (c) A megajoule is 1,000,000 joules and has the abbreviation MJ. (d) A nanonewton is $\frac{1}{1{,}000{,}000{,}000}$th of a newton and is abbreviated nN. (e) A centimeter is $\frac{1}{100}$th of a meter and is abbreviated cm.

Converting from one unit to the other is as simple as moving the decimal point and adding zeros to place the digits in the proper columns. For example, to convert from km to m, the decimal place is moved three places over:

1.54 km into m:

$$1.54 \quad = \quad 1540.\ m$$

where the final zero was added to make sure the decimal point is in the right place. 1.54 km is the same as 1540 m. This makes sense: meters is the smaller unit, so there are more of them in the same distance.

Moving the decimal point in the other direction:

1.54 mg into g:

$$1.54 \quad = \quad 0.00154\ g$$

where again, zeros have been added to make sure the digits are in the proper column. Notice that the decimal point is moved over one position for every zero in the scaling factor. There are three zeros in "1000 ×" and "$\frac{1}{1000}$th", so in both cases the decimal point is moved over three positions. If the scaling factor is a number greater than 1 (as 1000 is in the first example), then the decimal point moves to the right. If the scaling factor is less than one (as $\frac{1}{1000}$ is in the second example), then the decimal point is moved to the left.

Example 1.4. Change the following quantities into the requested units. (a) 0.154 km into m. (b) 8450 grams into kg. (c) 0.000 65 L into mL. (d) 0.000 65 L into μL. (e) 65,000,000 J into MJ. (f) 24.9 GW into W. (g) 0.000 000 03 g into μg.

Solutions. (a) 0.154 km is 154 m. (b) 8450 grams is 8.45 kg. (c) 0.000 65 L is 0.65 mL. (d) 0.000 65 L is 650 μL. (e) 65,000,000 J is 65 MJ. (f) 24.9 GW is 24,900,000,000 W. (g) 0.000 000 03 g is 0.03 μg.

You should become very familiar with the units, the prefixes, and how the prefixes are combined with the units. You should also become familiar with the abbreviations that are used and how they are combined, because it is very common in chemistry to use the prefixes and their abbreviations when using units.

Scientific Notation

The last two conversions in the previous example show that some numbers can get very, very large, and some numbers can get very, very small. (When we say "very small," we mean getting closer and closer to zero. We do not mean negative numbers.) It becomes troublesome to write all of the zeros necessary just to position one or two digits in the correct column. Since chemistry occasionally deals with very large or very small numbers, an easier way to express such numbers is needed.

Scientific notation is a simpler way of writing very large or very small numbers. It is based on powers of 10:

$$1{,}000{,}000{,}000 = 10 \times 10 \times 10 \times 10 \times 10 \times 10 \times 10 \times 10 \times 10 = 10^9$$
$$100{,}000{,}000 = 10 \times 10 \times 10 \times 10 \times 10 \times 10 \times 10 \times 10 = 10^8$$
$$10{,}000{,}000 = 10 \times 10 \times 10 \times 10 \times 10 \times 10 \times 10 = 10^7$$
$$1{,}000{,}000 = 10 \times 10 \times 10 \times 10 \times 10 \times 10 = 10^6$$
$$100{,}000 = 10 \times 10 \times 10 \times 10 \times 10 = 10^5$$
$$10{,}000 = 10 \times 10 \times 10 \times 10 = 10^4$$
$$1{,}000 = 10 \times 10 \times 10 = 10^3$$
$$100 = 10 \times 10 = 10^2$$
$$10 = 10 = 10^1$$

Large numbers are therefore positive powers of 10. The power or exponent is written as a small superscript on the right-hand side of the 10. The power in the numbers above is equal to the number of zeros following the digit 1. Ten raised to the 0 power is, by definition, 1. (Anything raised to the 0 power is 1.) Numbers smaller than one can be described by using negative powers of ten:

$$1 = 10^0$$
$$0.1 = \frac{1}{10} = 10^{-1}$$
$$0.01 = \frac{1}{10 \times 10} = 10^{-2}$$
$$0.001 = \frac{1}{10 \times 10 \times 10} = 10^{-3}$$
$$0.000\,1 = \frac{1}{10 \times 10 \times 10 \times 10} = 10^{-4}$$
$$0.000\,01 = \frac{1}{10 \times 10 \times 10 \times 10 \times 10} = 10^{-5}$$
$$0.000\,001 = \frac{1}{10 \times 10 \times 10 \times 10 \times 10 \times 10} = 10^{-6}$$
$$0.000\,000\,1 = \frac{1}{10 \times 10 \times 10 \times 10 \times 10 \times 10 \times 10} = 10^{-7}$$
$$0.000\,000\,01 = \frac{1}{10 \times 10 \times 10 \times 10 \times 10 \times 10 \times 10 \times 10} = 10^{-8}$$

The spaces between the zeros helps keep track of them more easily, like the commas do for the very large numbers. In this case, the negative exponent equals the number of zeros in the original number, **including the zero to the left the decimal point**.

Very large or very small numbers can be written using scientific notation. For example, 65,000,000 can be written as $6.5 \times 10,000,000$, which is 6.5×10^7 in scientific notation. We also see that 0.000 000 052 can be written as $5.2 \times 0.000\,000\,01$, which is 5.2×10^{-8}.

In writing down a number in scientific notation, the convention is to have a non-zero digit first, followed by a decimal point, then the rest of the non-zero digits, stopping with the last non-zero digit. Then, add the "times" sign, and write down 10 raised to the appropriate power. So, for example, while writing 59,220,000,000 in scientific notation as

$$59.22 \times 10^9$$

is technically not *in*correct, but it does not follow the accepted convention of a single non-zero digit before the decimal point. It is more conventional to write it as

$$5.922 \times 10^{10}$$

in proper scientific notation. By the same token,

$$0.5922 \times 10^{11}$$

while referring to the same number, is also not written according to convention. While it might seem that such conventions are nitpicking, understand that when we all follow the same conventions, it is much easier to communicate ideas and facts. A demand to follow a convention is not unique to chemistry, or even science. All fields have their own particular conventions that allow experts in that field to communicate with other experts more efficiently.

Example 1.5. Express the following numbers in appropriate scientific notation. (a) 66,900,000 (b) 0.005 83 (c) 12,001 (d) −0.000 082 07 (e) −3,141,000

Solutions. (a) 66,900,000 can be written as $6.69 \times 10,000,000$. Therefore, in scientific notation, the number is written 6.69×10^7 (b) 0.005 83 can be written as 5.83×0.001. Therefore, in scientific notation, this number is written as 5.83×10^{-3} (c) 1.2001×10^4 (d) -8.207×10^{-5} (e) -3.141×10^6.

Notice in the last two how the negative sign remains with the number itself, and does not affect the sign of the power.

Example 1.6. Change the units on the following quantities to the requested units, and express the final answer in proper scientific notation. (a) 0.000 45 g into µg. (b) 39.6 kJ into J. (c) 52 m into nm.

Solutions. (a) In going from grams to micrograms, the decimal point is moved six places to the right, so the answer in micrograms is 450 µg. In scientific notation, this would be 4.5×10^2 µg. (b) A kilojoule is 1000 joules, so 39.6 kJ is 39,600 J. In scientific notation, that would be 3.96×10^4 J. (c) There are 1,000,000,000 nm, or 10^9 nm, in a meter, so for 52 meters there are 52×10^9 nm. However, for proper scientific notation, this needs to be changed to 5.2×10^{10} nm. We got this answer by recalling that $52 = 5.2 \times 10^1$, and $10^1 \times 10^9 = 10^{10}$.

---

http://owl.thomsonlearning.com: Ch 0-2b Math: Exponential Notation

Scientific Notation and Calculators

These days, proper use of a calculator is practically mandatory in a chemistry class. "Proper" use, however, demands that you actually <u>know</u> how to work the calculator! You might be surprised to know how many beginning chemistry students **think** they know how to use their calculator properly but actually don't. This section, and several others in the chapters that follow, are devoted to proper use of calculators with respect to some of the topics discussed above. In particular, we need to review how numbers expressed in scientific notation are entered into a calculator.

Because different models of calculators are different, this discussion might not be applicable to ALL calculators. However, it will probably be applicable to almost all calculators. You should have a manual for your calculator, which you should consult for specific directions. However, even if your calculator doesn't work exactly like the ones discussed here, you can probably use these directions and, with your own manual, figure out how they are applicable to your own calculator.

The key to understanding how to enter a number using scientific notation is that **your calculator understands powers of 10**. What you enter into the calculator are the decimal part of the scientific notation expression, called the <u>mantissa</u>, and the power on the 10, which we have called the <u>exponent</u>:

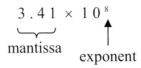

$$3.41 \times 10^{8}$$

mantissa   exponent

Calculators understand the "times 10 to the" part. All you need to do is enter into the calculator the 3.41 and the 8 correctly. The calculator then understands that you have entered in a number equal to 341,000,000.

When you enter a number into a calculator, it assumes you are entering a decimal-type number – unless you indicate otherwise by pushing a special key. Most calculators have a special key to indicate that the next numbers are the exponent. Look on your calculator for a key that looks like **EE** or **EXP** or, occasionally, **10ˣ**. You might have to invoke a second-function key to access it. This is the exponent function. When you press it, the next numbers that you enter are the value of the exponent. If you hit the **+/–** key, you negate the <u>exponent</u>. Most calculators only take two or three numbers into the exponent part of the number, along with a minus sign. Also, most calculators do not explicitly show a positive sign if the exponent is positive. Finally, many calculators do not show the "× 10" part of the scientific notation.

For example, suppose we want to enter 3.41×10^8 into a calculator. First, press the **3**, the decimal point **.**, the **4**, and the **1** keys on your keypad. The display of your calculator should look something like this:

3.4 1

Now, press the **EE** or **EXP** or **10ˣ** key to activate the exponent field. Your calculator display might look like something like this:

3.4 1   00

where the two zeros will represent the exponent. (Again, keep in mind that your particular calculator might have a different display. You should check your owner's manual if you are uncertain.) **Before touching any other calculator key, enter the exponent value.** You do not have to press the **X** key to indicate multiplication, or even enter 10. The calculator understands that it is 3.41 "times ten to the" correct power. After pressing the **8** key, your calculator display should now look like

3.4 1   08

If you need to negate the power – say, the number is 3.41×10^{-8} – you need to hit the **+/-** calculator key before you hit any other calculator key that performs an operation (like **X**, **÷**, or even **=**). Hitting the **+/-** key at this point would thus give you

$$3.4\,1 \qquad {}^{-08}$$

Notice that the negative sign is on the exponent, not the mantissa! You still have a positive number, but it's now equal to 0.000 000 034 1.

At this point, pressing any calculator key that performs an operation – **X**, **÷**, or even **=** – brings the calculator function out of the exponent field, and you can continue your calculation. Again, notice that we did not press **X**, nor did we enter 10. The calculator **understands** that the mantissa, 3.41, is being multiplied by 10 raised to a certain power, in this case 8. You, the calculator user, need to understand also how the calculator works. This becomes especially important when you start doing mathematical evaluations using your calculator.

If the number itself – that is, the mantissa – is negative, you have to negate the number **before** you hit your exponent key. Remember, when you activate the exponent portion of the display, any **+/-** changes the sign on the **exponent**, not the mantissa.

Example 1.7. State the number that is displayed in the following simulated calculator displays, and rewrite them in complete scientific notation and decimal form.

(a) $\qquad 5.8\,2\,5 \quad {}^{03}$

(b) $\qquad -9.7\,9\,4 \quad {}^{07}$

(c) $\qquad -1.0\,0\,3\,6 \quad {}^{-03}$

Solutions. (a) This number is 5.825×10^3, which would be written as 5825. (b) This number is a negative number, -9.794×10^7, which could also be written as $-97,940,000$. (c) This number has a negative exponent: it is -1.0036×10^{-3}, or $-0.001\,003\,6$.

Student Exercises

As with any skill, it is important that you (a) understand what you are doing and why, and (b) practice. The following exercises are meant to provide you with practice. Answers are given at the end of the chapter, but in order for you to find out if you truly understand, it is important that you try before seeing the answer! Some space is provided if you want to work in the book directly.

1.1 A patient has a systolic blood pressure of 120 millimeters of mercury. What is the complete unit of this measurement?

1.2. Identify the following units as either fundamental or derived:

(a) centimeter

(b) liter

(c) Kelvin

(d) newton

(e) joule

(f) second

1.3. Change 9.38 km into units of meters, and then change the meters into μm. Express your final answer in scientific notation.

1.4. Change the following quantities into the required units.

(a) 3.09×10^{-3} kW into W.

(b) 9.312×10^5 μs into s.

(c) 6.69×10^{-22} Gm into nm.

(d) 5 L into cL.

(e) 350 K into °C.

(f) 100°C into K.

1.5. Express the following numbers in proper scientific notation.

(a) 93,000,000

(b) 299,790,000

(c) 0.000 000 000 052 9

(d) 602,000,000,000,000,000,000,000

(e) 22.4

(f) 0.000 000 000 000 000 000 000 013 81

1.6. Express the following numbers in decimal form.

(a) 9.65×10^4

(b) 1.09677×10^5

(c) 2.38029×10^2

(d) 2.13×10^5

(e) -8.815×10^{12}

(f) 1.33758×10^{-3}

1.7. Using your calculator, properly enter the following three numbers in scientific notation. When you are done with a number, hit the = key and compare your calculator display to the simulated display in the answer section.

(a) 8.36×10^{12}

(b) 9.104×10^{-25}

(c) -7.772×10^{-9}

Answers to Student Exercises

1.1. The complete unit is "millimeters of mercury".

1.2. (a) derived (b) derived (c) fundamental (d) derived (e) derived (f) fundamental

1.3. 9.38 km = 9380 m = 9,380,000,000 μm = 9.38×10^9 μm.

1.4. (a) 3.09 W (b) 0.931 2 s (c) 0.000 669 nm or 6.69×10^{-4} nm (d) 500 cL (e) 77°C (f) 373 K.

1.5. (a) 9.3×10^7 (b) 2.9979×10^8 (c) 5.29×10^{-11} (d) 6.02×10^{23} (e) 1.381×10^{-23}.

1.6. (a) 96,500 (b) 109,677 (c) 238.029 (d) 213,000 (e) –8,815,000,000,000 (f) 0.001 337 58

1.7.

(a)	$8.3\,6^{\;12}$
(b)	$9.1\,0\,4^{\;-25}$
(c)	$-7.7\,7\,2^{\;-09}$

Chapter 2. Arithmetic Evaluation

Introduction

In the course of applying math to chemistry (or any science, for that matter), you will be asked to evaluate mathematical expressions. In many of such tasks, you will have to perform one or more of the four basic arithmetic operations: addition, subtraction, multiplication, and division.

Occasionally, you will have to deal with logarithms and exponentials, but most of the math you will perform will use the four primary operations. Some of these calculations can be done by hand. Many of them will be done using your calculator. Most calculators do not deal with units, so you will have to work those out yourself. This chapter reviews the methods that math has for evaluating various arithmetic expressions.

Addition and Subtraction

Addition is the combination of two quantities. Subtraction is the difference between two quantities. Addition is indicated by the plus sign (+), while subtraction is indicated by the minus sign (-). Addition and subtraction are considered mathematical opposites of each other. Subtraction can be thought of as the addition of a negative number:

$$100 - 23$$
$$= 100 + (-23)$$
$$= 77$$

One key to performing successful addition and subtraction in chemistry is to understand that **all of the quantities being added or subtracted must have the same unit.** Not only must they be the same type of the unit, but they must be the exact same unit. For example, it is obvious that one cannot add 2 kilometers and 5 grams. But some people add 2 kilometers to 5 meters and get 7 . . . kilometers? meters?

In order to add or subtract two quantities, their units must be the same. This may require that you change some of the units:

$$
\left.
\begin{array}{r}
5 \text{ kilometers} = 5000 \text{ meters} \\
+ \qquad 7 \text{ meters}
\end{array}
\right\} \text{ same unit}
$$
$$
\overline{\text{TOTAL} = 5007 \text{ meters}}
$$

If you wanted to, you could change the 7 meters to kilometers instead:

$$
\left.
\begin{array}{r}
7 \text{ meters} = 0.007 \text{ kilometers} \\
+ \ 5 \text{ kilometers}
\end{array}
\right\} \text{ same unit}
$$
$$
\overline{\text{TOTAL} = 5.007 \text{ kilometers}}
$$

and now the sum of the two numbers has units of kilometers. Which answer is correct? THEY BOTH ARE. Since 5.007 kilometers is equal to 5007 meters, both answers are correct. They simply have different units. Unless a problem specifically asks for a particular unit in the final answer, you should be able to choose which appropriate unit you want to determine the answer.

The above example applies to subtractions, too.

Example 2.1. Evaluate the following expression using two different units. (a) 6.8 grams + 0.8215 kilograms. (b) 310 K – 15°C. (For simplicity, we will use the approximate conversion K = °C + 273.)

Solutions. (a) If we want to evaluate the sum in units of grams, we will have to change 0.8215 kilograms into grams. Moving the decimal point over three places, it becomes 821.5 grams. Now we can perform the sum:

$$6.8 \text{ grams} + 821.5 \text{ grams} = 828.3 \text{ grams}$$

If we want to perform the sum using units of kilograms, we have to change 6.8 grams into kilograms. Moving the decimal point over three places, it becomes 0.0068 kilograms. Adding the two quantities:

$$0.0068 \text{ kilograms} + 0.8215 \text{ kilograms} = 0.8283 \text{ kilograms}.$$

You should satisfy yourself that the two answers are in fact the same, only with different units.

(b) We can express a final answer for the difference in the two temperatures in units of Kelvin or degrees Celsius. To convert 15°C into Kelvin,

$$15 + 273 = 288 \text{ K}$$

so that the difference becomes

$$310 \text{ K} - 288 \text{ K} = 22 \text{ K}$$

To express the answer in the other unit, we need to change 310 K to degrees Celsius:

$$310 - 273 = 37°\text{C}$$

Therefore,

$$37°\text{C} - 15°\text{C} = 22°\text{C}$$

The difference has the same value in both Kelvin and degrees Celsius units. That's because the degree size is the same for both scales. (This problem does not imply that 22°C of temperature equals 22 K of temperature! It means that a 22°C <u>change</u> in temperature is the same as a 22 K <u>change</u> in temperature.)

One of the most common subtractions you will perform in chemistry is to calculate the change in some measurement. Part (b) in the above example is a problem of this type. To determine the change in some measurement, **you always take the final value and subtract from it the initial value**:

$$CHANGE = FINAL\ VALUE - INITIAL\ VALUE$$

Sometimes the symbol Δ is used to indicate the change in some measurement. For example, ΔT would stand for the change in temperature, which is defined as the final temperature minus the initial temperature. Δmass represents the change in mass, which is the final mass minus the initial mass. Changes may have positive or negative values. If the change is positive, it means that the measurement is <u>increasing</u> in value. If the change is negative, then the measurement is <u>decreasing</u> in value.

Example 2.2. Determine (a) the change in temperature if an object goes from 50°C to 12°C, and (b) the change in density from 1.28 g/L to 3.72 g/L.

Solutions. (a) ΔT is equal to the final temperature, 12°C, minus the initial temperature, 50°C. Both measurements have the same units, so we can subtract directly:

$$\Delta T = 12°C - 50°C$$
$$\Delta T = -38°C$$

Notice that the change is negative. This means that the temperature is decreasing.

(b) Δdensity is equal to the final density, 3.72 g/L, minus the initial density, 1.28 g/L. Again, the densities have the same units, so we can evaluate the subtraction directly:

$$\Delta density = 3.72\ g/L - 1.28\ g/L$$
$$\Delta density = 2.44\ g/L$$

The change in density is positive, so in this case the density is increasing.

If you performed the above subtractions on your calculator, you should have gotten a negative sign in (a), telling you that the change was negative. Do not forget to include the negative sign as part of your answer!

Multiplication and Division: Numbers

By far, the majority of calculations you will perform in chemistry will involve multiplication and division. You might even perform both in the same calculation.

Recall that multiplication is a short-hand form of addition. Instead of adding 10 to itself four times

$$10 + 10 + 10 + 10 + 10 = 50$$

we say that we are taking 10 five times, and get 50. We write it

$$10 \times 5 = 50$$

which, reading aloud, says "ten times five equals fifty."

Division is the inverse of multiplication. Instead of taking a number many times, you are splitting it into smaller pieces. For example, if we want to separate 50 into five equal parts to get ten, we would write it

$$50 \div 5 = 10$$

which reads "fifty divided by five equals 10."

Instead of using the times sign, $\times$, the act of multiplication is usually indicated differently in some expressions. Sometimes, a simple dot, "·" is used to signify multiplication. For example, writing $7 \cdot 33$ implies the multiplication of 7 and 33. Even simpler, if numbers or variables are simply written next to each other, sometimes in parentheses, **multiplication is implied**. Therefore,

(7)(–66) implies the multiplication of 7 and –66, and the expression "*nRT*" implies the multiplication of the variables *n*, *R*, and *T*.

Division can also be written differently. Division is commonly represented with a <u>fraction</u>, which simply writes one number above another number, separated by a horizontal line. The number on the top, the <u>numerator</u>, is being divided by the number on the bottom, the <u>denominator</u>. For example, the division of 50 by 5 is written as

$$\frac{50}{5} = 10$$

which is still read as "fifty divided by five equals 10." In this case, 50 is the numerator and 5 is the denominator. A fraction may also be written in-line, like 50/5 – which equals 10.

A useful trick (one which we will use in Chapter 4) is to recognize that any number can be thought of as a numerator **divided by one**. For example, the number 10 can be thought of as $\frac{10}{1}$.

With that in mind, we define the <u>reciprocal</u> of a number as the number made by switching the numerator and denominator of any fraction. The reciprocal of $\frac{2}{3}$ is $\frac{3}{2}$:

$$\frac{2}{3} \quad = \quad \frac{3}{2}$$

Because any number can be thought of as being a fraction with one in the denominator, any number has a reciprocal. The reciprocal of 10 is $\frac{1}{10}$:

$$10 = \frac{10}{1} \qquad \text{reciprocal} = \frac{1}{10}$$

Multiplication and division can be considered reciprocal operations of each other. Multiplying and dividing a quantity by the same number leaves the original quantity unchanged. For example,

$$3.5 \times 6.7 \div 6.7 = 3.5$$

Another way of writing this is by using a fraction:

$$\frac{3.5 \times 6.7}{6.7} = 3.5$$

The 6.7 in the numerator and denominator cancel each other.

We will occasionally take advantage of the fact that **multiplication by a number is the same as division by the reciprocal of that number. Division by a number is the same as multiplication by the reciprocal of that number.** Sometimes we can simplify expressions by rewriting them with reciprocals. For example:

$$66 \times \frac{1}{4} = \underbrace{66 \div 4}_{\substack{\text{Easier to plug} \\ \text{into calculator?}}} = \frac{66}{4} = 16.5$$

In this expression, it may be easier for some students to see how to plug numbers into a calculator to evaluate the multiplication operation as a division by the reciprocal. The same tactic can be used when there are fractions within numerators or denominators:

$$\frac{6/7}{3/4} = \frac{6}{7} \div \frac{3}{4} = \frac{6}{7} \times \frac{4}{3} = \frac{24}{21} = \frac{8}{7}$$

where in the center step we are using the fact that multiplication with the reciprocal is equivalent to division.

Example 2.3. Evaluate the following expressions. (a) $6.32 \div \dfrac{8}{14}$ (b) $9 \cdot 8 \cdot 7 \div 5$.

Solutions. (a) $6.32 \div \dfrac{8}{14}$ is the same thing as $6.32 \times \dfrac{14}{8}$, which is 11.06. (b) We can rewrite the expression so that it looks like a fraction: $9 \cdot 8 \cdot 7 \div 5 = \dfrac{9 \cdot 8 \cdot 7}{5}$, which equals 100.8.

One of the nice things about multiplication and division is that they are <u>commutative</u>. This means that if you have an expression where you have to multiply and divide numbers, the exact order in which you perform the functions does not matter. Consider Example 2.3 (b), above. There are several ways we can do the multiplications and divisions:

$$\frac{(9 \cdot 8 \cdot 7)}{5} = \frac{504}{5} = 100.8$$

$$\frac{9}{5} \cdot 8 \cdot 7 = 1.8 \cdot 7 \cdot 8 = 1.8 \cdot 56 = 100.8$$

$$9 \cdot \frac{8}{5} \cdot 7 = 9 \cdot 1.6 \cdot 8 = 100.8$$

and many other possibilities. However, all of these combinations give the same, correct answer! Therefore, when evaluating expressions with many numbers in the numerator and denominator, the order of operation does not matter, as long as only multiplication and division are being performed. We will consider order of operations in more detail later in this chapter.

This is true, of course, only as long as you evaluate the expression properly! With many calculators, there can be a potential problem. Consider the following combination of multiplication and division:

$$\frac{4.6 \cdot 63.9}{2.55 \cdot 34.0}$$

Try entering this into your calculator. If you got 3.3903...., then you evaluated the expression properly. However, some people will get 3919.199......, which is incorrect. What happened?

For people who get the incorrect answer, this is the usual pattern of keystrokes on their calculator:

$$\textbf{4.6} \quad \textbf{X} \quad \textbf{63.9} \quad \div \quad \textbf{2.55} \quad \textbf{X} \quad \textbf{34.0} \quad \textbf{=}$$

because, of course, that's what the expression is. But what many people don't realize is that on most calculators, every time you hit an operation key (i.e. **X, ÷, =**, etc.) the calculator evaluates the entire expression up to that point! Therefore, when that second **X** key was entered, the calculator evaluated the entire expression up to that point, which was

$$\frac{4.6 \cdot 63.9}{2.55} \text{, which} = 115.2705....$$

Then, when the second **X** key was hit, the calculator assumes that this number, 115.2705...., was being multiplied by 34.0, **which the calculator assumes is in the numerator!**

Because many calculators evaluate expressions whenever an operation button is pressed, the proper way to evaluate the expression would be to either use your parentheses keys:

$$\textbf{4.6} \quad \textbf{X} \quad \textbf{63.9} \quad \div \quad \textbf{(} \quad \textbf{2.55} \quad \textbf{X} \quad \textbf{34.0} \quad \textbf{)} \quad \textbf{=}$$

Or, you can use the ÷ operation instead of the **X** operation for the final number:

$$\textbf{4.6} \quad \textbf{X} \quad \textbf{63.9} \quad \div \quad \textbf{2.55} \quad \div \quad \textbf{34.0} \quad \textbf{=}$$

Either way should give you the proper answer.

Some more advanced calculators will not evaluate an expression until the **=** key is pressed. Usually, these calculators have large viewscreens that show all of the numbers and operations so you can check your expression before hitting **=**. Many of these calculators also have alphanumeric capability; i.e. you can enter words as well as numbers. If your calculator is like this, much of the above discussion may not apply strictly. You should read your calculator manual to determine what the correct order of entry is to evaluate expressions like the ones above. That's the basic rule for calculator use: **know how to use your calculator!**

Multiplication and Division: Units

Like numbers, we can multiply and divide units too. In fact, we have already seen some examples of that in the last chapter, when we discussed derived units. Units of volume are found using (length) × (length) × (length), and so we have units like m^3 and cm^3 for volume. We call them "meters cubed" or "cubic meter" and "cubic centimeter." When different units are multiplying together, as in N·m, we speak it as "newton-meter." As with numbers, the assumption here is that unless otherwise stated, derived units are assumed to be <u>products</u> of the more basic units. Thus, when you hear "kilowatt-hours," it is correct to understand that this unit is "kilowatts <u>times</u> hours."

Units can also be divided. Perhaps one of the most common examples of this in everyday life is the unit for velocity, miles per hour. Mathematically, the word "per" implies a division, so miles per hour can be written in fraction form as

$$\frac{\text{miles}}{\text{hour}}$$

Similarly, if a unit is given as a simple fraction, it can be stated using the word "per" to indicate the division. Thus,

$$\frac{\text{grams}}{\text{mole}}$$

can be stated as "grams per mole."

Derived units can get much more complicated than this, as suggested by the table of derived units given in Chapter 1. An important point about units, however, is that they follow the same algebraic rules of factoring out of numerators and denominators that numbers do. If the exact same unit is in the numerator and denominator, it can be crossed out of both.

If, for example, you have the following combination of units,

$$\frac{km}{sec} \cdot \frac{sec}{hr}$$

then the rules of fractions state that the product of theses two fractions is the product of the numerator quantities divided by the product of the denominator quantities. Thus, this equals

$$\frac{km \cdot sec}{sec \cdot hr}$$

But the unit "sec" appears in both numerator and denominator, so it can be canceled out:

$$\frac{km \cdot \cancel{sec}}{\cancel{sec} \cdot hr} = \frac{km}{hr}$$

In Chapter 4, we will use this idea a lot.

Quite a few derived units may not make much sense at first glance. For example, we are relatively familiar with the derived unit $\frac{km}{hr}$, kilometer per hour, as a unit of velocity. Another example would be $\frac{g}{mL}$ as a unit of density. But some derived units may not make as much sense up front. For example, the unit $\frac{kg \cdot m}{sec^2}$ does not have immediate physical significance to most of us. Neither does $\frac{L \cdot atm}{mol \cdot K}$, which is one possible set of units for the ideal gas law constant, a very important quantity in chemistry.

The reason for these seemingly complex units is <u>mathematical</u> <u>necessity</u>. Recall that **units must follow the same rules of algebra as numbers do**. This means that the algebraic combination of units must be performed along with the algebraic combination of numbers. This requires us to take certain, seemingly unusual, mathematical combinations of seemingly unrelated units.

For example, consider Newton's second law: if a force, F, acts on a body, the magnitude of the force is proportional to the acceleration, a, caused by the force as well as to the mass of the body, m. Mathematically, this is written as $F = ma$. Mass has units of kg, and acceleration has units of m/sec/sec, which can be shown to be equal to $\frac{m}{sec^2}$. (You can show this by applying the rules of reciprocals from the previous section to units, not just numbers!) Therefore, when multiplying a mass time an acceleration, you also have to multiply their <u>units</u>: $kg \cdot \frac{m}{sec^2}$, or $\frac{kg \cdot m}{sec^2}$. This means that force must have units of $\frac{kg \cdot m}{sec^2}$. It does, and for simplicity's sake we rename this derived unit the <u>newton</u> and give it the symbol N. Ultimately we use the newton as the unit of force, but we must always remember that it is originally defined as $\frac{kg \cdot m}{sec^2}$.

Not all derived units like this get renamed. For example, $\frac{L \cdot atm}{mol \cdot K}$, which we will use in ideal gas problems, is not given a new name. However, the algebra we use to manipulate these units, as we will see in Chapter 6, requires that certain quantities have such unusual derived units.

Example 2.4. Determine the overall unit for the final answer that you will get by performing the following operations. (a) Mass is multiplied by velocity to give a quantity called momentum. (b) Electric charge, which has units of coulombs, C, is squared and then divided by distance to get energy. (c) Current, which has units of amps, is multiplied by time in seconds to get units of electric charge in units of coulombs, C.

Solutions. (a) With mass having units of kg and velocity having units of $\frac{m}{sec}$, momentum has units of $kg \cdot \frac{m}{sec}$ or $\frac{kg \cdot m}{sec}$. (b) Squaring coulombs gives C^2, and dividing that by m, the unit of distance, gives $\frac{C^2}{m}$ as a unit of energy. [NOTE: This derived unit is not equal to J, the SI unit of energy. However, there is a conversion factor. Check your textbook to see if it deals with these topics.]

(c) Current times time gives units of amp·sec, which according to the example is a definition of the unit coulomb: C = amp·sec.

One very important consideration is the fact that **most calculators do not work with units, only with numbers**. It is up to you, the student, to interpret the way the units in a problem are handled. Sometimes this takes a little practice. However, in the long run you will find that a proper handling of the units in these algebraic problems is not only necessary, but can sometimes actually assist you in figuring out how to work a problem. Chapters 4, 5, and 6 will give many examples of how we use units to help work a problem.

Algebra and Equations

A portion of the mathematical skills you will be performing in chemistry is using a given equation to solve for some unknown quantity. These equations are composed of variables whose meanings, at least, you should be familiar with. In all cases, you will probably be asked to solve for one of the variables when all of the other variables will be known to you (or you can find them out).

An <u>equation</u> is any mathematical expression that contains an equals sign, =. We have already seen some very simple forms of equations, like

$$\frac{50}{5} = 10$$

An equation implies that what is on the left side of the = sign is the same as what is on the right side of the = sign. In the above example, we recognize that the fraction $\frac{50}{5}$ reduces to $\frac{10}{1}$ which equals 10, so the overall equation reduces to "10 = 10," which we know to be true. On the other hand, if we had the equation

$$\frac{55}{5} \overset{?}{=} 10$$

then we know that something is wrong because $\dfrac{55}{5}$ does not equal 10. This is not a proper equation.

Many equations are composed of variables, not numbers. <u>Definitions</u> are among the simplest of equations of variables. For example, the definition of average velocity can be written as:

$$\text{average velocity} = \frac{\text{distance traveled, in m}}{\text{time taken, in sec}}$$

Since the quantity on the left must equal the quantity on the right, if we have distance and time, we can determine the average velocity. If the distance were 400 m and it took 10 seconds to travel that distance, then

$$\text{average velocity} = \frac{400\,\text{m}}{10\,\text{sec}} = \frac{400}{10}\frac{\text{m}}{\text{sec}} = 40\frac{\text{m}}{\text{sec}}$$

where we are showing that $40\,\dfrac{\text{m}}{\text{sec}}$ is equal to $\dfrac{400\,\text{m}}{10\,\text{sec}}$.

One of the things that you must be able to do is to solve for <u>any</u> variable in an equation, if you are given all of the others. If your average velocity were 25 meters per second and you needed to travel 5000 meters (which is five kilometers), how many seconds would that take you?

Since the quantities involved are the same as used for the definition of average velocity, **you can use the same equation to determine an answer.** However, you know different quantities – and you are seeking to solve for a different quantity, also. What you have is

$$25\frac{\text{m}}{\text{sec}} = \frac{5000\,\text{m}}{\text{time taken}}$$

and you have to solve for the time, in units of seconds.

There are two keys to being able to solve for any variable in an equation:

➢ The quantity you are looking for **must be all by itself on one side of the equation** (it doesn't matter which side);

➢ The quantity you are looking for **must be in the numerator**; that is, you should be able to write the quantity as a value divided by one in the denominator.

Performing mostly multiplication and division, you can rearrange any equation to isolate the desired quantity by itself in the numerator; then simply evaluate the numbers and units on the other side of the equation to get your final answer.

For the question above, the quantity you are looking for (sometimes referred to as "the unknown" in an equation) is "time taken." You need to rewrite the equation algebraically to isolate that quantity by itself in the numerator on one side of the equation. (It's in the denominator right now.) There are several ways to do this algebraically. What follows is not the only way. But, if you perform all of the algebra properly (no matter what way you do it), you should get the same answer.

First, we will multiply both sides of the equation by the variable "time taken." We can do this because, when starting with an equality, if you perform the same operation to both sides of the equality, it is still an equality. (The only exception is multiplying or dividing by zero.) We get

$$25\frac{m}{sec} \cdot \text{time taken} = \frac{5000 \text{ m}}{\text{time taken}} \cdot \text{time taken}$$

Now consider the right side of the equation. Because we have "time taken" in the numerator and the denominator of the fraction, they can cancel out:

$$25\frac{m}{sec} \cdot \text{time taken} = \frac{5000 \text{ m}}{\cancel{\text{time taken}}} \cdot \cancel{\text{time taken}}$$

$$25\frac{m}{sec} \cdot \text{time taken} = 5000 \text{ m}$$

Now, we divide both sides of the equation by $25 \frac{m}{sec}$. Again, we can do this to both sides of the equation and still have an equality:

$$\frac{25\frac{m}{sec} \cdot \text{time taken}}{25\frac{m}{sec}} = \frac{5000\ m}{25\frac{m}{sec}}$$

Notice that we are including the units along with the numbers. On the left side, the $25\ \frac{m}{sec}$ is in the numerator and denominator, and it cancels out:

$$\frac{25\cancel{\frac{m}{sec}} \cdot \text{time taken}}{25\cancel{\frac{m}{sec}}} = \frac{5000\ m}{25\frac{m}{sec}}$$

We have canceled the 25 as well as the units $\frac{m}{sec}$! What we have left is

$$\text{time taken} = \frac{5000\ m}{25\frac{m}{sec}} = \frac{5000}{25}\ \frac{m}{\frac{m}{sec}}$$

where now we have what we are looking for, the time taken, all by itself on one side of the equation and in the numerator. (Remember, it can be thought of as a fraction with one in the denominator.) Using our calculator, we evaluate the numerical part of the answer on the right-hand side: $\frac{5000}{25} = 200$. To evaluate the units, which most calculators will not do, we recall that a division can also be considered as multiplication by the reciprocal:

$$\frac{m}{\frac{m}{sec}} = m \div \frac{m}{sec} = m \times \frac{sec}{m} = sec$$

where in the last step, the meter unit has been canceled out from the numerator and the denominator. The final unit on the answer is sec, so the complete answer is

$$\text{time taken} = 200 \text{ seconds}$$

One thing to keep in mind with regard to units. **They make sense**. We were looking for an amount of time, and the answer has a unit of time, <u>seconds</u>. If we performed the units analysis (sometimes called <u>dimensional</u> <u>analysis</u>) incorrectly and got meters for our answer, we would say that the time taken was 200 meters. HUH? That doesn't make sense! "Meters" is a unit of length, not a unit of time. **Always ask yourself if the unit you get for an answer is consistent with the quantity you are looking for**. In this way, keeping track of units can actually help you work out a solution to a problem.

Example 2.5. The density of a material is defined as the mass of the material divided by its volume. (a) Write the mathematical equation for the definition of density. (b) What is the density of mercury if 24 milliliters has a mass of 326.4 grams? Express the density in those units. (c) How many milliliters of osmium are needed to have a mass of 100 grams if its density is 22.4 grams per milliliter? (d) What is the mass of 1.1×10^6 L of hydrogen gas if it has a density of 0.0899 grams per liter?

Solutions. (a) Using the definition of average velocity as a guide, we see that the mathematical definition of density can be written as

$$\text{density} = \frac{\text{mass}}{\text{volume}}$$

(b) The density of mercury is found by plugging into the definition of density:

$$\text{density} = \frac{\text{mass}}{\text{volume}} = \frac{326.4 \text{ g}}{24 \text{ mL}} = 13.6 \frac{\text{g}}{\text{mL}}$$

(c) In order to determine the volume of osmium necessary, we set up the density equation as

$$22.4\,\frac{g}{mL} = \frac{100\ g}{volume}$$

The manipulations of this equation are similar to the velocity example from above: multiply both sides by volume, then divide both sides by the value of the density. We get:

$$volume = \frac{100\ g}{22.4\,\dfrac{g}{mL}} \approx 4.46\ mL$$

where we use the "$\approx$" sign to mean "approximately equal to," since we have limited our answer to three digits. We check our final unit and recognize that mL is in fact a unit of volume.

(d) The mass of hydrogen for the volume given takes only one rearrangement step:

$$0.0899\,\frac{g}{L} = \frac{mass}{1.1 \times 10^{6}\,L} \quad \Rightarrow \quad 1.1 \times 10^{6}\,L \cdot 0.0899\,\frac{g}{L} = mass$$

Multiplying the two quantities together and canceling out the L unit, we find that the mass = 98,890 grams of hydrogen, or just under 99 kg.

http://owl.thomsonlearning.com: Ch 0-3a Math: Basic Algebra

Order of Operations

Addition, subtraction, multiplication, division, exponents and logarithms, etc., are called operations. They ask that you do something to the numbers or expressions on either side of the operation sign (i.e. $+$, $-$, X. $\div$, log, etc.), which are called operators.

When trying to figure out an expression that has a combination of operators, you cannot simply evaluate them in any order. For example, the following expression contains addition and multiplication (in terms of an exponent):

$$(2 + 3)^2$$

This expression cannot be evaluated properly by distributing the square through the parentheses, squaring the numbers, and adding them together. It is **incorrect** to evaluate this expression this way:

$$(2 + 3)^2 = (2^2 + 3^2) = 4 + 9 = 13 \quad \text{WRONG!}$$

This is incorrect. Exponents are not distributed through an expression in parentheses. In this case, the square requires that we evaluate

$$(2 + 3)^2 = (2 + 3)(2 + 3) = (5)(5) = 25$$

which is the correct way to evaluate the original expression.

The rules of algebra are set up so that a particular order of evaluating operations is required to evaluate a complicated expression correctly. The correct order of operations is

- ➢ **Parentheticals** – evaluate the expressions inside parentheses
- ➢ **Raised powers** – perform any exponential operations (or inverse exponential operations, like square roots and logs)
- ➢ **Multiplication & Division** – perform any products or quotients
- ➢ **Addition & Subtraction** – combine numbers or variables together by adding or subtracting

Notice that Multiplication & Division and Addition & Subtraction are grouped together in the same steps. That's because algebra recognizes the inverse relationship between these operations. In order to remember the order Parentheticals, Raised powers, Multiplication & Division, and Addition & Subtraction, some people use the mnemonic "Please Remember My Dear Aunt Sally". The first letter in each word, PRMDAS, stands for the type of operation. Notice that while this mnemonic device puts division after multiplication and subtraction after addition, division and multiplication are actually evaluated together, since they are in reality inverse operations. Addition and subtraction are also actually evaluated together, since they too are really inverse operations.

It is important to realize that in some complicated expressions, you will have an expression inside an expression inside an expression . . . etc. These are called <u>nested</u> expressions. The rule for the order of evaluating nested expressions is to **evaluate the innermost expression first and work your way to the outermost expression in steps**.

It is not often that you will have to deal with all of the algebraic operations in any one problem. It is difficult, then, to come up with relevant examples that illustrate all of the possibilities. Hopefully, the few given here will suffice to give you the general idea.

To evaluate the expression

$$K = \frac{(2.0)^2(3.0)}{(0.1)^2(0.5)^3}$$

we first evaluate the raised power for each number inside each set of parentheses. $(2.0)^2 = 4.0$, $(0.1)^2 = 0.01$, and $(0.5)^3 = 0.125$. We now have

$$K = \frac{4.0 \cdot 3.0}{0.01 \cdot 0.125}$$

Now we can evaluate the multiplication and division operations to get

$$K = 9600 = 9.6 \times 10^3$$

as the correct answer. You might want to try plugging this into your calculator to verify that this is indeed the correct numerical answer. Suppose we have a more complicated expression, like

$$E = 0.46 - \frac{(8.314)(298)}{(2)(96,500)} \cdot \ln\left(\frac{(1)(2)^3}{(0.5)^2}\right)$$

There are several parts to this one expression, which need to be worked out before everything can be brought together. (This is why it's nice to have a calculator that has a memory function, which you should know how to use properly.) The "ln" is the natural logarithm, and the expression asks that you take the natural logarithm of an expression. (Logarithms will be covered in more detail in Chapter 7.) This is an example of nested expressions. The following diagram shows the order we will take to evaluate the entire expression:

$$E = 0.46 - \underbrace{\underbrace{\frac{(8.314)(298)}{(2)(96,500)}}_{\#3} \cdot \ln \underbrace{\left(\frac{\overbrace{(1)(2)^2}^{\#1}}{(0.5)^2} \right)}_{\#2}}_{\#4}$$
$$\underbrace{\phantom{0.46 - \frac{(8.314)(298)}{(2)(96,500)} \cdot \ln \left(\frac{(1)(2)^2}{(0.5)^2} \right)}}_{\#5}$$

Step #1 evaluates the nested expression inside the logarithm. Evaluating the raised powers and then performing the multiplication and division (which is the correct order of operations), we get

$$\left(\frac{(1)(2)^2}{(0.5)^2} \right) = \frac{1 \cdot 8}{0.25} = 32$$

Step #2 evaluates the natural logarithm of the inside expression, which equals 32. On most calculators, this is performed by entering **32** and then pressing the natural logarithm key, which usually looks like **ln** or **LN**. On a few calculators, like high-powered engineering or graphing calculators, you might have to enter **ln 32** and then press the **=** or **EXECUTE** key. (Check your calculator manual for instructions about how to take logarithms properly using your particular model.) You should get

$$\ln 32 = 3.4657...$$

We will round off to three digits and approximate this as 3.47. Step #3 calls for the evaluation of the fraction in front of the logarithm. If you perform this properly using your calculator, you should get

$$\frac{(8.314)(298)}{(2)(96,500)} = 0.0128...$$

Step #4 has us multiplying this expression by the logarithm term. Notice that this is still a multiplication, so we are on the same step in our PRMDAS scheme.

$$0.0128 \cdot 3.47 = 0.04436$$

and the final step, #5, has us evaluating the subtraction:

$$E = 0.46 - 0.04436 = 0.41564$$

which we will limit to two decimal places to get $E = 0.42$. (We have not considered units in this example. This was intentional.) Please note that this example followed the conventions of nested expressions and the proper order of operations, PRMDAS, as required. When you develop the right level of sophistication after enough practice, you might be able to evaluate an expression like this using your calculator in one long set of keystrokes. But, don't force it: it is better to evaluate complex expressions in steps, correctly, than to try and do it in one long step – but get the incorrect final answer!

Student Exercises

Space is provided for you to work out the problems here in the book. Answers to all student exercises are given at the end of the exercises, on the next page. But give each problem your best effort before you check yourself!

2.1. Evaluate the following expressions twice, using two different units.

(a) 4.50 kg + 870 g

(b) 923 mg + 2.980 g

(c) 124 cm + 23.00 m

(d) 1.775 L – 43 mL

(e) 350 K – (–43°C)

2.2 (a) Write 45 – 12 as an addition and evaluate.

(b) Write –700 + (–375) as a subtraction and evaluate.

(c) Write 650 + 350 as a subtraction and evaluate.

2.3. (a) What is the change in position of a football player going from the 10-yard line to the 50-yard line?

(b) What is the change in temperature of a cup of coffee going from 75°C to 12 °C?

(c) What is your change in altitude going from +8850 m (above sea level) to –77 m (below sea level)?

(d) What is the change in mass if a plant grows from 70.4 g to 160.3 g?

2.4. Write the following expressions as fractions and evaluate them.

(a) $6.6 \times 3.0 \div 0.18$

(b) $(99 \times -3.6) \div (22.07 \times 0.007)$

(c) $1 \div (467 + 33)$

(d) $1 + (1 \div (1 + (1 \div (1 + 1))))$

2.5   Write the reciprocal expressions for 2.4 (a) – (c) and evaluate them. Without writing the reciprocal expression, what is the value of the reciprocal of 2.4 (d)?

2.6 (a) Rewrite the expression $6 \div \dfrac{2}{3}$ as a multiplication and evaluate.

(b) Rewrite the expression $\dfrac{22}{7} \times \dfrac{1}{11}$ as a division and evaluate.

2.7. Does $2 \times 3 \times 4 \times 5$ equal $5 \times 4 \times 3 \times 2$? Why or why not?

2.8. Use your calculator to evaluate the following expressions without writing anything down.

(a) $\dfrac{3}{4 \cdot 5}$ (b) $\dfrac{(9.5)(-44.2)}{50.4}$ (c) $\dfrac{(0.005\,56)(4567)}{(39.06)(120.6)}$ (d) $\dfrac{(23+95)(54-98)}{(87)(21)}$

2.9. What units should you get when you:

(a) divide units of time by units of velocity?

(b) multiply J·sec by $\dfrac{1}{\text{sec}}$?

(c) divide newtons by $\dfrac{\text{m}}{\text{sec}^2}$? (HINT: What are the units that make up the newton unit?)

2.10. How long will it take to travel 50 km at an average velocity of 25 meters per second?

2.11. Evaluate the following expressions. (a) $88 - \dfrac{(2.5 + 7.7)^2}{-3.8}$ (b) $\sqrt{\dfrac{67 - (18 + 33)}{-\left(1 - \left(\dfrac{4}{0.0069}\right)\right)}}$

(c) $(2.9)(44)(-0.072) \div (11.8 - 5)$.

2.12. Evaluate the following expressions.

(a) $\dfrac{(2)(0.08205)(298)}{0.987}$

(b) $100 \cdot \sqrt{\dfrac{1}{1 - \left(\dfrac{2 \times 10^8}{3 \times 10^8}\right)^2}}$

(c) $-0.76 - \dfrac{8.314 \cdot 298}{3 \cdot 96,500} \ln \dfrac{(0.75)^2(0.50)}{(1.75)^3(1.05)}$

2.13. Is either addition or subtraction commutative? Why or why not?

Answers to Student Exercises

2.1. (a) 5.37 kg, 5370 g (b) 3.903 g, 3903 mg (c) 24.24 m, 2424 cm (d) 1.732 L, 1732 mL (e) 120°C, 120 K.

2.2. (a) 45 + (−12), which equals 33. (b) −700 − (+375) or −700 − 375, which equals −1075. (c) 650 − (-350), which equals 1000.

2.3. (a) 40 yards (b) −63°C (c) −8927 m (d) 89.9 g.

2.4. (a) 110 (b) −2306.9.... (c) 0.002 (d) 1.666 66.....

2.5. The values for the reciprocals are (a) 0.009 090.... (b) −0.000 433 47... (c) 500 (d) 0.6.

2.6. (a) 9 (b) $\dfrac{2}{7}$.

2.7. Yes, because multiplication is commutative; it doesn't matter what order the values are multiplied in.

2.8. You should get (a) 0.15 (b) −8.3313.... (c) 0.005 390... or 5.390... × 10^{-3} (d) −2.8418...

2.9. (a) You get units of distance, like m. (b) J (c) kg.

2.10. 2000 seconds, or 0.556 hours.

2.11 (a) 115.37..... (b) 0.1662... (c) 1.351....

2.12. (a) 49.54... (b) 134.16.... (c) 0.734...

2.13. Addition is commutative if all of the numbers you are combining are added together, because the order the numbers are added in does not matter. Subtraction, however, is not commutative. As a simple example, consider 2 − 1 and 1 − 2. They do not equal the same number. However, addition of negative numbers is commutative. For example, 2 + (−1) is equal to (−1) + 2.

place would b
significant figι
uncertain to (u

Example 3.1.

(a) a pressure
(Burettes are t
line. Units of ν
Solutions. (a)
"65" line. We
position of the
the pressure as
marking the b
represents 46.5
However, it is
estimate the liι
we have 4 foι
significant figu

Deterι

Often y
to determine s

Introduction

Perhaps one of the most unusual concepts that beginning chemistry students need to come to grips with is the idea of significant figures, or "sig figs." Significant figures arise in two situations: in doing mathematical calculations and in making measurements in, say, a chemistry lab. The best way to illustrate the need to follow the rules for assigning sig figs is by example. Suppose you are given that 23 grams of potassium perchlorate, $KClO_4$, has a volume of 9.1 mL. You are asked to find the density. You plug the numbers into the density formula:

$$\text{density} = \frac{\text{mass}}{\text{volume}} = \frac{23 \text{ grams}}{9.1 \text{ mL}}$$

Satisfied that you set up the problem correctly, you use your calculator and evaluate the fraction $\frac{23}{9.1}$ and, reading the numbers from the calculator, get 2.52747252747.... $\frac{g}{mL}$. Now ask yourself: does it make sense to include all of these digits when your original numbers, 23 and 9.1, only had two digits each? No, it does not, and the concept of significant figures addresses that.

Suppose you are asked to make a measurement, like reading the temperature in Celsius from the thermometer below:

where the thick line represents the mercury in the thermometer. Since there are ten divisions between each number (count them and see), each division represents a tenth of a degree, 0.1°C. Well, you know that the temperature is at least 33.0°, and the mercury is just beyond the line that would represent 0.2, so the temperature is at least 33.2°C. You can guess that the edge is about a third of the way to 0.3, so you might estimate that the edge is reading about 0.23, so the overall temperature is about 33.23°C. How about estimating the thousandths' place in the temperature so

number of significant figures. Therefore we limit our final answer to 3 significant figures. Our final answer, after the proper rounding, is

$$V = 23.3$$

Example 3.5. Evaluate the following expressions, limiting the final answer to the proper number of significant figures. (a) $(68.3)(0.003097)(2)$; the 2 is exact. (b) $-\dfrac{(8.314)(298.15)}{96,500}$ (c) $(0.0650)(6210)$ (d) $\dfrac{223.0}{71.0}$ (This is sometimes used as an approximation for π.)

Solutions. (a) The calculator answer is 0.4230502, but we have to limit it to three significant figures, so our final answer is 0.423. (b) The calculator answer is $-2.56872446 \times 10^{-2}$, but we must limit the answer to three significant figures. Therefore, the final answer is -2.57×10^{-2}. Notice that we had to round up, and that the 10^{-2} does not affect, or is not affected by, the significant figure determination. (c) The calculator answer is 403.65, but we limit the answer to three significant figures and get 404. (d) The calculator gives 3.14084507042.... as the answer, but we need to limit the answer to three significant figures, so we give 3.14 as the final answer.

What about problems that have addition and subtraction **and** multiplication and division? Probably the best tactic is to determine the proper number of significant figures for each part of the problem that you evaluate, using the appropriate rules. For example, suppose you need to evaluate the expression

$$4.32 - 56.92 \times (22.87 - 22.73)$$

Keeping in mind the general rule "PRMDAS," we will evaluate the expression inside the parentheses first, then perform the multiplication, then subtract that number from 4.32. So, for the first step, we evaluate

$$(22.87 - 22.73) = 0.14$$

Notice that this subtraction took us from four significant figures to only two! According to the order of operations, now the multiplication is evaluated. This means we must multiply 56.92 by the evaluated parenthetical expression.

$$56.92 \times 0.14 = 7.9688 = 8.0$$

We have limited 7.9688 to two significant figures, as required. Finally, we subtract 8.0 from 4.32.

$$4.32 - 8.0 = -3.68 = -3.7$$

We limit the significant figures to the tenths' place, since this is the rightmost common significant digit for the two numbers. Our final answer is –3.7, which has two significant figures.

Usually, you can keep all non-significant figures until your final answer. This means that you should not apply rules for sig figs when you perform the multiplication and division until after the addition and subtraction are performed. Your final answer should be very close to the answer, if not even the same answer, that you get if you limit your intermediate answers to the proper significant figures. If, for example, we performed all of the above operations and then applied the strictest sig fig rule (in this case, limitation of two sig figs from the multiplication operation), we would get

$$4.32 - 56.92 \times (22.87 - 22.73) = -3.6488 = -3.6$$

as our final answer. Notice that this differs from our earlier by one in the last significant digit. This is an example of <u>truncation error</u>, which is the slight variation of your final answer caused by imposing limits on the number of significant figures at different points in the calculation. Is this a big problem? No, it isn't. Remember that sig figs are those digits that are known, **plus the first uncertain digit.** Truncation errors are a natural part of significant figures, and their existence has influenced the rules by which significant figures are determined.

Significant figures are really an issue of common sense. Don't report more significant digits, numbers that are supposed to mean something, beyond the ability of the measurement device or beyond the significance of the numbers you use to do your calculations. We made a point in earlier material that most calculators do not work out the units. Well, most calculators don't care about significant figures, either. It's up to the people working with the numbers to decide how many of the digits are significant.

http://owl.thomsonlearning.com: Ch 0-2a Math: Significant Figures, Questions 4, 5, 6

Student Exercises

3.1. How long is each rod? Assume the units are centimeters.

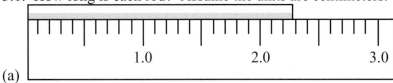

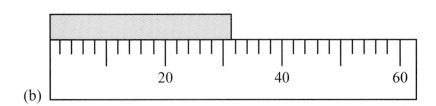

(a)

(b)

3.2. Read the following gauges to the proper number of significant figures.

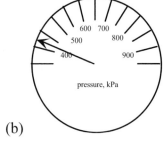

(a)   (b)   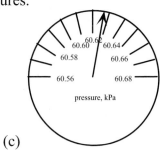(c)

3.3. How many significant figures are in the following numbers? (a) 5.009×10^5 (b) 0.006510
(c) 1.000001 (d) –9.01.

3.4. How many significant figures are in the following numbers? (a) 0.00005 (b) 978,000
(c) 6.022×10^{23} (d) $6.6260755 \times 10^{-34}$.

3.5. Evaluate the following expressions to the proper number of significant figures. (a) 55,993 +
354.9 (b) 24,080 – 35 (c) 0.00103 – 0.00088 (d) $7.22 \times 10^3 + 6.89 \times 10^2$

3.6. Evaluate the following expressions to the proper number of significant figures. (a) 102,993 + 6700 + 12,065 (b) 0.9634 – 0.0622 – 0.029 (c) 7845 + 3460 – 22.4

3.7. Would you consider π an exact number or not? Explain your answer.

3.8. Evaluate the following expressions to the proper number of significant figures. (a) 22.4 × 8.314 × 298.15 (b) $\dfrac{4.184 \cdot 2.08}{1.987}$ (c) $2 \cdot \dfrac{(1.61)^2}{4 \cdot \pi \cdot (6.626)^2}$ (assume that the 2, 4, and the π are exact).

3.9. Given the following expression

$$\frac{(745)(V)}{310.5} = \frac{(802)(4.50)}{298.15}$$

solve for V to the correct number of significant figures.

3.10. Given the following expression

$$\frac{(603.65)(25.88)}{598.2} = \frac{(499.20)(12.0)}{T}$$

solve for T to the correct number of significant figures.

3.11. Given the following expression

$$\Delta G = -(3)(96,477)(-1.019)$$

solve for ΔG to the correct number of significant figures. The 3 is exact.

3.12. Evaluate the following expressions. Impose the rules for significant figures different ways for the same expression. Do you find any truncation errors?

(a) $99.0 \times (44.1 - (-22.007))$

(b) $0.633 \div 21.9 + (3.40 \times 0.09022)$

(c) $1.0 \div 1.0 + (1.0 \div 7.5)$

(d) $\dfrac{(681)(432 + 98.22)}{(-1005)(2.00 \div 0.00446)}$

Answers to Student Exercises

3.1. (a) About 2.28 cm long. (b) About 31.2 cm long.

3.2. (a) About 5.42 kPa. (b) About 420 kPa. (c) About 60.628 kPa

3.3. (a) 4 sig figs (b) 4 sig figs (c) 7 sig figs (d) 3 sig figs.

3.4. (a) 1 sig fig (b) 3 sig figs (c) 4 sig figs (d) 8 sig figs.

3.5. (a) 56,348 (b) 24,050 (rounding up) (c) 0.00015 (d) 7.91×10^3, or 7910.

3.6. (a) 121,800 (b) 0.872 (c) 11,280.

3.7. If all of the digits of π were used in a calculation, it should be considered as an exact number. If you only use a few digits, it is not exact. Simply using π as a variable implies it is exact, but using 3.14 as an approximation makes it not exact.

3.8. (a) 55,500 (b) 4.38 (c) 9.40×10^{-3}.

3.9. 5.04

3.10. 229

3.11. 294,900, or 2.949×10^5

3.12. (a) 6540 (b) 0.336 (c) 1.1 (d) –0.802 if sig fig limits applied at each step in parentheses, –0.801 if sig fig rules (limit to three figures) applied only at the end of calculation.

Chapter 4. Converting Units

Introduction

In many chemistry calculations, you will be required to change units from one kind to another. This action is called several different things, and your book may prefer one name over another: dimensional analysis, the factor-label method, or, the preference here, converting units. Unit conversion is one of the most useful and widely applicable mathematical tools. In fact, we will extend this idea in the next chapter and see just how powerful a method it is.

Making Fractions from Equalities

We will start with the idea that 1 meter is equal to 100 centimeters. That is the definition of centimeter. We can write this statement as an equation:

$$1 \text{ meter} = 100 \text{ centimeter}$$

$$1 \text{ m} = 100 \text{ cm}$$

where in the second equation, we are using the abbreviations of the units. We should have an intuitive understanding that 100 cm **is** 1 m, so the above mathematical equation is acceptable to us.

Now, remember the rule in math that says that if you do something (i.e. add, subtract, multiply, divide) to both sides of an equation, what you generate is still an equation. That is, the left side still equals the right side of the equation. You have to perform the same operation **on both sides**. In this case, let us divide both sides of the equation by 100 centimeters. (And why not? There's no mathematical rule that forbids us to divide an expression with something that has a unit.) We get

$$\frac{1 \text{ m}}{100 \text{ cm}} = \frac{100 \text{ cm}}{100 \text{ cm}}$$

On the right-hand side of the equation, we have the same thing in the numerator and the denominator of the fraction, so it cancels. Not only the number cancels, but the unit cancels as well. Remember that we treat units algebraically, just like we treat numbers:

$$\frac{1 \text{ m}}{100 \text{ cm}} = \frac{100 \text{ cm}}{100 \text{ cm}} = 1$$

Because everything in the numerator and denominator cancels, we are left with 1. Therefore,

$$\frac{1 \text{ m}}{100 \text{ cm}} = 1$$

The value of the fraction $\frac{1 \text{ m}}{100 \text{ cm}}$ is exactly 1. This might not be as intuitively obvious as our original equation (1 m = 100 cm), but it makes sense. One meter **is** 100 centimeters, and a fraction that has the same quantity in the numerator and denominator is equal to one. It's just that in this fraction, the numerator and denominator are expressed in different units.

Keep in mind that we can also divide the original equation by 1 meter instead of 100 centimeters:

$$1 \text{ m} = 100 \text{ cm}$$

$$\frac{1 \text{ m}}{1 \text{ m}} = \frac{100 \text{ cm}}{1 \text{ m}}$$

$$\frac{1 \text{ m}}{1 \text{ m}} = \frac{100 \text{ cm}}{1 \text{ m}} = 1$$

$$\frac{100 \text{ cm}}{1 \text{ m}} = 1$$

and we get another fraction that equals one. You should be able to see that $\frac{100 \text{ cm}}{1 \text{ m}}$ is the reciprocal of $\frac{1 \text{ m}}{100 \text{ cm}}$, and the reciprocal of 1 is still 1. We therefore have two expressions that equal 1.

What we need to keep in mind is that **when you multiply a quantity by 1, that quantity does not change.**   For example, 5×1 is still 5, and 25 meters times 1 is still 25 meters.   The absolute value of the quantity does not change.

However, we can use the fractions that are equal to one to convert the units of an amount of something into different units.   The absolute quantity remains the same, but the units we use to express that quantity are changed.   We do this by expressing an equivalence as a fraction, putting the unit we want **from**, and its associated number, in the **denominator** of the fraction.   The unit (and its associated number) we want to convert **to** is in the numerator.   We multiply this fraction, which equals 1, by our initial quantity.   Then, we multiply and divide our numbers and get a new value with a new unit – but it's still the same quantity.

For example, we know that 50 centimeters is half of a meter, or 0.5 meters.   Mathematically, however, we can set up a conversion like this:

$$50 \text{ cm} \cdot \frac{1 \text{ m}}{100 \text{ cm}}$$

where the fraction $\dfrac{1 \text{ m}}{100 \text{ cm}}$ equals 1, and so by multiplying 50 cm by this fraction, we are making a new expression that is equivalent to the original quantity "50 cm."   Remember that we can treat the 50 cm as a numerator with 1 in the denominator.   What this means is that we have the product of two fractions **with the unit of cm in both numerator and denominator.**   Therefore, the unit "cm" cancels:

$$50 \, \cancel{\text{cm}} \cdot \frac{1 \text{ m}}{100 \, \cancel{\text{cm}}}$$

When we perform the remaining multiplications and divisions, we get

$$50 \cdot \frac{1 \text{ m}}{100} = 0.50 \text{ m}$$

where our final answer now has units of meters.   The quantity has not changed its absolute **value**, it has only changed its **units**.

There are two simple hints for converting units:

➢ **Start with something that you are given.** In a majority of mathematical problems, this rule works. In problems, you will be given some data, some quantities. In most cases, the problem can be solved by taking one of the quantities and converting it to another unit.

➢ **The unit you want to get rid of usually goes in the denominator of the conversion factor you multiply the quantity by.** In doing this, we use the normal rules of algebra and cancel units from numerator and denominator.

As a counterexample, suppose we were to multiply our 50 cm by the other fraction that equals 1, the $\dfrac{100 \text{ cm}}{1 \text{ m}}$:

$$50 \text{ cm} \cdot \frac{100 \text{ cm}}{1 \text{ m}}$$

This is OK, isn't it? After all, we are only multiplying our 50 cm by 1 again. But in this case, the **units** don't work out: the cm unit does not cancel algebraically, and we get a nonsensical expression if we try to multiply and divide everything through. **Use the units to help you construct proper conversion factors and it will work out correctly every time.** If you don't pay attention to the units in these conversion factors, you will quickly go wrong. Keeping track of the units can actually help you work out a problem.

Example 4.1. Set up and evaluate the following conversions algebraically. (a) 0.75 m to cm. (b) 3.66 km to m. (c) 1855 mL to L. (d) 3.88×10^4 g to kg.

Solutions. The following are the algebraic constructions for performing the conversions. Notice that they all start with the quantity that is given. You might want to plug the numbers into your calculator to verify the final numerical result, and write on the page to show how the units cancel.

(a) $0.75 \text{ m} \cdot \dfrac{100 \text{ cm}}{1 \text{ m}} = 75 \text{ cm}$ (b) $3.66 \text{ km} \cdot \dfrac{1000 \text{ m}}{1 \text{ km}} = 3660 \text{ m}$

(c) $1855 \text{ mL} \cdot \dfrac{1 \text{ L}}{1000 \text{ mL}} = 1.855 \text{ L}$ (d) $3.88 \times 10^4 \text{ g} \cdot \dfrac{1 \text{ kg}}{1000 \text{ g}} = 3.88 \times 10^1 \text{ kg} = 38.8 \text{ kg}$

More Than One Conversion

Of course, more than one conversion can be applied to a quantity. In that case, each conversion can be performed sequentially. For instance, in converting 0.0678 km into millimeters, we will first convert the km units into units of meters:

$$0.0678 \; \cancel{km} \cdot \frac{1000 \text{ m}}{1 \; \cancel{km}} = 67.8 \text{ m}$$

where the kilometer units in the numerator and denominator are canceled algebraically. We now convert the meters units into millimeters,

$$67.8 \; \cancel{m} \cdot \frac{1000 \text{ mm}}{1 \; \cancel{m}} = 67,800 \text{ mm}$$

canceling the meter units, for our final answer. Any number of proper conversions can be performed sequentially like this.

Of course, we could have performed this conversion in a single step if we recognized that there are 1,000,000 mm in 1 km. But instead of memorizing all possible conversion factors between the prefixed units (like how many microliters there are in a gigaliter), **it is usually easier to first convert from the the original prefixed unit to the basic (unprefixed) unit, and from the basic unit to the final prefixed unit.** In the above case, we went from kilometers (a prefixed unit, with the prefix being "kilo-") to meters (the basic unit) and then from meters to millimeters (the final prefixed unit, with "milli-" being the prefix this time).

Example 4.2. Perform the following conversions in two steps by first converting to the basic, unprefixed unit in each case. (a) 5.7×10^{-5} kW into μW (W = watts) (b) 9.992×10^{12} mHz into MHz (where the unit hertz, Hz, is a unit of frequency and equals $\frac{1}{\text{sec}}$).

Solutions. (a) To go from kilowatts to microwatts, we first convert our units into the basic unit of watts,

$$5.7 \times 10^{-5} \; \cancel{kW} \cdot \frac{1000 \text{ W}}{1 \; \cancel{kW}} = 5.7 \times 10^{-2} \text{ W}$$

and then convert the watts into microwatts,

$$5.7 \times 10^{-2} \; \cancel{W} \cdot \frac{10^{6} \; \mu\text{W}}{1 \; \cancel{W}} = 5.7 \times 10^{4} \; \mu\text{W}$$

Our final answer is 5.7×10^4 µW. For (b), it is a similar process except we are converting millihertz into megahertz. The first step is to convert the millihertz to the basic unit of hertz,

$$9.992 \times 10^{12} \text{ mHz} \cdot \frac{1 \text{ Hz}}{1000 \text{ mHz}} = 9.992 \times 10^9 \text{ Hz}$$

and now from Hz to MHz,

$$9.992 \times 10^9 \text{ Hz} \cdot \frac{1 \text{ MHz}}{10^6 \text{ Hz}} = 9.992 \times 10^3 \text{ MHz} = 9992 \text{ MHz}$$

In both cases, the unit cancellations are shown.

After performing a lot of unit conversions, you might begin to feel that they are all very similar, and they are. The same mathematical techniques are used for all unit conversions, so when you master those techniques, you can do any conversion problem.

After a bit of sophistication in these conversion problems is developed, we can take the process to another level. Instead of performing multi-step conversions in separate parts, **we can group the steps into a single, longer step.** Mathematically, our final answer will be exactly the same as if we performed the conversion in steps – as long as we include all of the proper steps.

We will convert 0.0678 km into mm again, but this time in one long step:

$$0.0678 \text{ km} \cdot \frac{1000 \text{ m}}{1 \text{ km}} \cdot \frac{1000 \text{ mm}}{1 \text{ m}} = 67,800 \text{ mm}$$

We get the same answer as before. Notice how the units work out: the first fraction cancels the "km" unit,

$$0.0678 \text{ km} \cdot \frac{1000 \text{ m}}{1 \text{ km}} \cdot \frac{1000 \text{ mm}}{1 \text{ m}} = 67,800 \text{ mm}$$

and the second fraction cancels the "m" unit,

$$0.0678 \text{ km} \cdot \frac{1000 \text{ m}}{1 \text{ km}} \cdot \frac{1000 \text{ mm}}{1 \text{ m}} = 67,800 \text{ mm}$$

The only unit left is millimeter, which is the unit we are converting into. In entering the numbers into a calculator, we could either hit the **=** or **ENTER** or **EXECUTE** key after each number, or just use the **X** key before entering any number in a numerator and the **÷** key before any number in a denominator. Our sequence of keys would then be

$$\textbf{0.0678 X 1000 X 1000 =}$$

(where we are not including the 1's in the denominators).

Finally, we recognize the mathematical equivalence of this method and doing the conversion in steps by pointing out that the first two terms, the 0.0678 km and the first fraction, are collectively **equal to 67.8 m**:

$$\underbrace{0.0678 \; \cancel{km} \cdot \frac{1000 \; \cancel{m}}{1 \; \cancel{km}}}_{= \; 67.8 \; m} \cdot \frac{1000 \; mm}{1 \; \cancel{m}} = 67{,}800 \; mm$$

In a sense, all we are doing is a substitution of the first step into the second step of this conversion. In doing so, we accomplish the complete conversion all at once instead of a step at a time.

Example 4.3. Re-do the two problems in Example 4.2, but in a single step.

Solutions. You should be able to convince yourself that the two examples will look like this:

(a) $5.7 \times 10^{-5} \; kW \cdot \dfrac{1000 \; W}{1 \; kW} \cdot \dfrac{10^{6} \; \mu W}{1 \; W} = 5.7 \times 10^{4} \; \mu W$

(b) $9.992 \times 10^{12} \; mHz \cdot \dfrac{1 \; Hz}{1000 \; mHz} \cdot \dfrac{1 \; MHz}{10^{6} \; Hz} = 9.992 \times 10^{3} \; MHz = 9992 \; MHz$

In both cases, we get the same answer. For (b), the sequence of keys in your calculator might be

$$\textbf{9.992 EE 12 ÷ 1000 = ÷ 1 EE 6 =}$$

The first **=** evaluates the division from the first conversion factor, and the very next **÷** sign indicates that the next number, which is really $\underline{1 \times 10^6}$, is also in the denominator and should be dividing the expression. **You should practice putting these numbers in your calculator and verify that you get the same answer given here.** If you do not, you may not be using your calculator correctly.

Density: A New Type of Conversion Factor

The <u>density</u> of an object is defined as the mass of the object divided by its volume:

$$\text{density} = d = \frac{\text{mass}}{\text{volume}}$$

Density usually has units of $\frac{g}{mL}$, $\frac{g}{cm^3}$, or even $\frac{g}{L}$ if the material is a gas. These units are read as "grams per milliliter," "grams per cubic centimeter," and "grams per liter," respectively. For example, the density of iron metal is $7.86 \ \frac{g}{cm^3}$.

Remember that fractions can be thought of as having a "1" in the denominator. Therefore, the density of iron can be written as

$$\frac{7.86 \ g}{1 \ cm^3}$$

which is read as "7.86 grams per one cubic centimeter." That is, for every cubic centimeter of volume, a piece of iron has a mass of 7.86 grams. Another way of writing this is as an equation. We say that, for iron,

$$7.86 \ g = 1 \ cm^3$$

Density, therefore, represents a conversion factor between mass and volume. If we have a mass, we can use density to determine the volume of that mass. If we have a volume, we can use density to determine the mass of that volume.

There are three things to keep in mind. First, different materials have different densities. That's why we specified that the above density is "for iron." The above density is for iron only. It would not be correct to use this value for the density if your sample is, say, water or balsa wood or sodium chloride. To keep this in mind, it may be better to write the mass-volume relationship above as

$$7.86 \text{ grams of iron} = 1 \text{ cm}^3 \text{ of iron}$$

to emphasize that this equivalence is applicable only to iron. (We will use memory devices like this a lot in the next chapter.) In working a problem, you must use the correct density for the material being considered. Second, your mass and volume units must be consistent when working out a mathematical problem! If your density is given in terms of $\frac{g}{mL}$ and your volume is in units of liters, you must convert a volume unit so that it will appropriately cancel out in order to determine the mass of your sample.

Third, recognize that you can use the above equation to make two different conversion factors. The first is the normal definition of density:

$$\frac{7.86 \text{ g of iron}}{1 \text{ cm}^3 \text{ of iron}}$$

This is useful for volume-to-mass conversions. You can **also** put the <u>mass</u> in the denominator:

$$\frac{1 \text{ cm}^3 \text{ of iron}}{7.86 \text{ g of iron}}$$

This is mathematically the reciprocal of the density, and can be used for mass-to-volume conversions. The density thus acts as two different conversion factors. Which one do you use? It depends on what unit you are trying to cancel and what unit you ultimately want for your final answer.

Example 4.4. These examples all use density as a conversion factor and go from simple to more complex. (a) What is the mass of 11.83 cm^3 of iron metal? (b) What is volume of 100.0 grams of silver nitrate, $AgNO_3$, which has a density of 4.35 grams per cubic centimeter? (c) The airship *Hindenberg*, which exploded in 1937, was filled with 2.000×10^{11} mL of H_2 gas. At a density of 0.0824 grams per liter, how many kilograms of hydrogen is this?

Solutions. (a) In order to convert from volume to mass, we use the density conversion factor that has volume in the denominator. This corresponds to the original definition of density:

$$11.83 \; \cancel{cm^3 \; of \; iron} \cdot \frac{7.86 \; grams \; of \; iron}{1 \; \cancel{cm^3 \; of \; iron}} = 93.0 \; grams \; of \; iron$$

where we limit our final answer to 3 significant figures. Notice how the "cm^3 of iron" units cancel. (b) To convert from mass to volume, we use the inverse of density because we want the units of mass to cancel:

$$100.0 \; \cancel{g \; of \; AgNO_3} \cdot \frac{1 \; cm^3 \; of \; AgNO_3}{4.35 \; \cancel{g \; of \; AgNO_3}} = 23.0 \; cm^3 \; of \; AgNO_3$$

Again, the final answer is limited to three significant figures. (c) In order to determine the mass of hydrogen, we must make the volume units consistent with each other. We can either convert the volume of H_2 to liters, or the density of H_2 to grams per mL. The answer will be the same either way the problem is worked. Let us convert the initial volume to units of liters, then use the density as a conversion factor to determine the number of grams of H_2 we have. Finally, we will include one more conversion factor to convert the mass in grams to a mass in units of kilograms:

$$2.000 \times 10^{11} \; mL \cdot \underbrace{\frac{1 \; L}{1000 \; mL}}_{converts \; to \; liters} \cdot \underbrace{\frac{0.0824 \; g}{1 \; L}}_{converts \; to \; mass} \cdot \underbrace{\frac{1 \; kg}{1000 \; g}}_{converts \; to \; kg} = 1.65 \times 10^4 \; kg$$

Each step in this problem, the most complicated one we have worked out so far, cancels a unit and introduces a new one. The final conversion gives us the unit we need, kg. You should practice plugging the numbers into your calculator for each example – especially the last one! – and verify that you can reproduce the answer that is given.

Other Conversion Factors

There are several other quantities defined in science and chemistry that are similar to density in that they contain two (or more) different types of units and therefore represent conversion factors between those types of units. Density is the conversion factor between mass and volume. Other quantities that can be used mathematically **just like density** are listed on the next page.

Quantity	Abbreviation	Definition	Types of Units
moles	mol	1 mole = 6.02×10^{23} atoms or molecules	amount and atoms/molecules
moles	mol	1 mole = 1 molecular weight of material, in grams	amount and mass
molarity	M	$\dfrac{\text{\# moles of solute}}{\text{\# L of solution}}$	moles and volume of solution
molality	m	$\dfrac{\text{\# moles of solute}}{\text{\# kg of solvent}}$	moles and mass of solvent
energy per mole	$\dfrac{kJ}{mol}$	$\dfrac{\text{amount of energy}}{\text{mole of material}}$	energy and amount
rate of reaction	$\dfrac{mol}{sec}, \dfrac{M}{sec}$	$\dfrac{\text{change in amount}}{\text{change in time}}$	amount and time

In all cases, the abbreviation represents a new unit that, with a number, can be used to express the related quantity. (For example, "0.25 M" is a concentration that is equal to 0.25 moles of solute per liter of solution.) For each quantity, either there is an equation in the definition that can be turned into a conversion factor (like the first two entries) or the definition is written as a fraction that, like density, can be used directly to go from one type of unit to another (i.e. the last four entries). In both cases, the quantities can be used algebraically just like we used density in the previous section. In the next chapter, we will begin to use these quantities quite a bit when we work out mathematical problems.

All of the new units above are related to <u>amount</u>. In chemistry, amount of material is a very important quantity and its unit, the <u>mole</u>, is perhaps the most useful unit of chemistry.

Converting Combined Units

Suppose we have a unit that is a combination of several fundamental units. How do we convert those? What you do is **break apart the combined unit into the product of its individual units and multiply by a conversion factor for each individual unit.**

For example, we have defined a newton as a $\dfrac{kg \cdot m}{sec^2}$. There is another unit of force called the erg, which has the fundamental unit definition of $\dfrac{g \cdot cm}{sec^2}$. How many ergs are there in one newton?

We apply a conversion factor for each unit that we need to change. First, let us convert the kg unit to g:

$$1 \frac{\cancel{kg} \cdot m}{sec^2} \cdot \frac{1000 \text{ g}}{1 \cancel{kg}} = 1000 \frac{g \cdot m}{sec^2}$$

Then, we can convert the m to cm (of course, the order of conversions does not matter; you could do m to cm first, then kg to g):

$$1000 \frac{g \cdot \cancel{m}}{sec^2} \cdot \frac{100 \text{ cm}}{\cancel{m}} = 100,000 \frac{g \cdot cm}{sec^2} = 10^5 \text{ erg}$$

There are 100,000 ergs in one newton. Similar conversions can be performed just as easily.

Example 4.5. The liter was originally defined as a cubic decimeter, dm^3. A decimeter is one-tenth of a meter. How many cubic centimeters, cm^3, are there in a liter?

Solution. This example requires several conversions. We will start with the fact that

$$1 \text{ L} = 1 \text{ dm}^3$$

and convert to units of meters, and then to units of centimeters. We will rewrite the cubic decimeter as the explicit product of three decimeter units:

$$1 \text{ dm}^3 = 1 \text{ dm} \cdot dm \cdot dm$$

It will be easier to see the conversion this way. Using the rules for converting combined units, we will need to convert each dm to m, which means that we will have to use the same conversion factor **three times**:

$$1 \cancel{dm} \cdot \cancel{dm} \cdot \cancel{dm} \cdot \frac{1 \text{ m}}{10 \cancel{dm}} \cdot \frac{1 \text{ m}}{10 \cancel{dm}} \cdot \frac{1 \text{ m}}{10 \cancel{dm}} = 0.001 \text{ m}^3$$

Now we can convert the m units to cm units by again writing the m³ as m·m·m and applying the appropriate conversion factor three times:

$$0.001 \; \cancel{m^3} \cdot \frac{100 \text{ cm}}{10 \; \cancel{m}} \cdot \frac{100 \text{ cm}}{10 \; \cancel{m}} \cdot \frac{100 \text{ cm}}{10 \; \cancel{m}} = 1000 \text{ cm}^3$$

There are 1000 cm³ in one liter. Since there are also 1000 mL in one liter, we establish the fact that

$$1 \text{ mL} = 1 \text{ cm}^3$$

where both 1's can be considered exact numbers (so they don't affect the significant figure determination).

Student Exercises

Try to express your final answers in the proper number of significant figures.

4.1. Construct two conversion factors from the following relationships.

(a) 1 inch = 2.54 cm

(b) 1 dozen = 12 things

(c) 1 m = 10^6 μm

(d) 1 cm³ = 1 mL

4.2. Convert, in a single step:

(a) 5.209 kW into W

(b) 20.0 μL into L

(c) 38,000 g into Mg

(d) 0.000 000 035 torr into mtorr (torr is a unit of pressure)

4.3. Convert, in two steps:

(a) 35.4 MHz to mHz

(b) 7.24×10^7 nm to km

(c) 29.11 μL to mL

(d) 0.95 msec to nsec

4.4. The human eye is most sensitive to green light, which has a wavelength of about 5500 Å, where 1 Å ('Ångstrom') equals 1×10^{-10} m. What is the wavelength of green light in units of m?

4.5. Evaluate the following expressions using your calculator, and determine the correct final unit:

(a) $78.3 \text{ kg} \cdot \dfrac{1000 \text{ g}}{1 \text{ kg}} \cdot \dfrac{1000 \text{ mg}}{1 \text{ g}} =$

(b) $0.073 \text{ L} \cdot \dfrac{832 \text{ g}}{1 \text{ L}} \cdot \dfrac{1 \text{ mol}}{40.04 \text{ g}} =$

(c) $1.088 \text{ L} \cdot \dfrac{1 \text{ mol}}{22.4 \text{ L}} \cdot \dfrac{28.02 \text{ g}}{1 \text{ mol}} \cdot \dfrac{1000 \text{ mg}}{1 \text{ g}} =$

4.6. What is the density of a material, in units of $\dfrac{\text{g}}{\text{cm}^3}$, if 0.497 kg has a volume of 0.107 L? HINT: Use the answer in Example 4.5 to help with your unit conversion.

4.7. What mass of helium has a volume of 125.0 L if the density of helium gas is 0.1787 $\frac{g}{L}$?

4.8. Sodium chloride, NaCl, has a density of 2.17 $\frac{g}{cm^3}$. If you needed 250.0 grams of NaCl, what volume would you need?

4.9. Which has more mass, 2750 mL of argon gas having a density of 1.784 $\frac{g}{L}$ or 4.50 cm^3 of phenol, which has a density of 1.058 $\frac{g}{cm^3}$? (Argon gas is sometimes used to fill light bulbs; phenol is the active ingredient in some sore-throat lozenges.)

4.10. How many mm^3 are there in one liter?

4.11. How many atoms are there in 1.00 grams of xenon gas? Xenon has at atomic weight of 131.3 $\frac{grams}{mole}$. HINT: You will have to use both definitions of the unit "mole" in a two-step conversion.

4.12. If a chemical reaction gives off 383.51 kJ of energy per mole of carbon dioxide produced as a product, how much energy is this in units of calories per gram of CO_2 produced? The molecular weight of CO_2 is 44.00 grams per mole, and there are 4.184 J in 1 calorie. (This is an exact conversion).

4.13. What volume of solution, in liters, is necessary to obtain 0.250 moles of solute from a solution whose concentration is given as 0.1059 M?

4.14. If you are given a sodium chloride solution that is 2.054 M in concentration, what volume of solution is needed to get 25.0 grams of dissolved NaCl? The formula weight of NaCl is 58.44 grams per mole.

Answers to Student Exercises

4.1. (a) $\dfrac{1 \text{ inch}}{2.54 \text{ cm}}$ and $\dfrac{2.54 \text{ cm}}{1 \text{ inch}}$ (b) $\dfrac{1 \text{ dozen}}{12 \text{ things}}$ and $\dfrac{12 \text{ things}}{1 \text{ dozen}}$ (c) $\dfrac{1 \text{ m}}{10^6 \text{ μm}}$ and $\dfrac{10^6 \text{ μm}}{1 \text{ m}}$

(d) $\dfrac{1 \text{ cm}^3}{1 \text{ mL}}$ and $\dfrac{1 \text{ mL}}{1 \text{ cm}^3}$

4.2. (a) 5209 W   (b) 0.000 020 0 L, or 2.00×10^{-5} L   (c) 0.038 Mg, or 3.8×10^{-2} Mg
(d) 0.000 035 mtorr, or 3.5×10^{-5} mtorr.

4.3. 35,400,000,000 mHz, or 3.54×10^{10} mHz   (b) 0.0724 km, or 7.24×10^{-2} km   (c) 0.02911 mL, or 2.911×10^{-2} mL   (d) 950 nsec, or 9.5×10^2 nsec.

4.4. 0.000 000 55 m, or 5.5×10^{-7} m.

4.5. (a) 78,300,000 mg, or 7.83×10^7 mg   (b) 1.5 mol   (c) 1360 mg, or 1.36×10^3 mg.

4.6. $4.64 \dfrac{\text{g}}{\text{cm}^3}$.

4.7. The helium would have a mass of 22.34 grams.

4.8. 115 cm^3 of NaCl would be needed.

4.9. The argon would have a mass of 4.91 grams, and the phenol would have a mass of 4.76 grams. The argon would therefore have slightly more mass.

4.10. There are 1,000,000 (or 1×10^6) mm^3 in one liter.

4.11. There are 4.58×10^{21} atoms of xenon in 1.000 gram of xenon.

4.12. The reaction gives off 2083 calories for every gram of CO_2 produced.

4.13. You will need 2.36 liters of solution to get 0.250 moles of solute.

4.14. You will need 0.208 liters of solution to get 25.0 grams of dissolved sodium chloride from the solution.

Chapter 5. Using Chemical Reactions to Make Conversion Factors

Introduction

One of the central themes in chemistry is the <u>balanced chemical reaction</u>. It provides, in a nutshell, a statement of what initial chemical substances are changing into what final chemical substances. Respectively, these substances are called <u>reactants</u> and <u>products</u>. Because of the Law of Conservation of Mass, the amount of mass of reactants must equal the amount of mass of products, and that's why we have to write **balanced** chemical reactions.

A balanced chemical reaction therefore contains **quantitative** information, both in terms of grams and moles of reactants and products. This quantitative information can be used to perform a variety of calculations. This chapter reviews some of the techniques for using balanced chemical reactions mathematically, by constructing conversion factors to use in calculations.

Balanced Chemical Reactions and Moles

Consider how chemistry summarizes the reaction of hydrogen gas, H_2, with oxygen gas, O_2, to make the product water, H_2O:

$$H_2 \ (g) \ + \ O_2 \ (g) \ \rightarrow \ H_2O \ (\ell)$$

We will drop the phase labels in the future, for reasons of clarity. The first step in writing any balanced chemical reaction is to start with proper formulas for the reactants and products. Both hydrogen and oxygen are diatomic elements, and the correct formula for water is well known.

If we consider this reaction at the atomic scale, then what we are saying is, "One molecule of hydrogen and one molecule of oxygen are reacting to make one molecule of water." This is the correct way of wording it, but there's a problem: it violates the Law of Conservation of Mass. If you count the number of hydrogen **atoms** as reactants, there are a total of two; and if you count the number of hydrogen atoms as products, there are also two. But there are two oxygen atoms as reactants and **only**

one oxygen atom in the products! Where did the other oxygen atom go? One of the fundamental ideas in chemistry is that you must have the same number of atoms of each element as products and as reactants.

We cannot change the formulas of the individual reactants and products. However, we can change the number of molecules of reactants and products. This takes some trial-and-error and experience, but in time it becomes a relatively easy task. Properly-balanced reactions typically use the lowest whole numbers necessary to make the number of atoms of each element the same on both sides of the reaction. For the reaction of hydrogen and oxygen to make water, the properly balanced reaction is

$$2\,H_2 \;+\; O_2 \;\rightarrow\; 2\,H_2O$$

You should check this to make sure there is the same number of atoms of each element on both sides of the reaction.

This balanced chemical reaction says, "Two molecules of hydrogen and one molecule of oxygen react to make two molecules of water." This is a proper interpretation of the balanced chemical reaction. But it is very difficult to follow reactions at the atomic or molecular scale. Atoms and molecules are just too small for us to follow them reacting one at a time. Chemistry defines a unit called the <u>mole</u> that solves this problem. A mole represents a certain number of things, just as a dozen represents a certain number of things. But whereas a dozen is 12, a mole represents a much larger number:

$$1\ \text{mole} = 6.02 \times 10^{23}$$

This number is called <u>Avogadro's</u> <u>number</u>, after an Italian chemist who proposed its existence. Why does a mole represent this number of things? Well, atoms have mass. In very small units called <u>atomic</u> <u>mass</u> <u>units</u> (or <u>amu</u>), one hydrogen atom has a mass of about 1.008 amu. One oxygen atom has a mass of about 16.00 amu, and an average mercury atom has a mass of about 200.6 amu. One **mole** of hydrogen atoms, 6.02×10^{23} H atoms, has the same amount of mass but in units of grams: one mole of hydrogen atoms has a mass of 1.008 **grams**. One mole of mercury atoms, 6.02×10^{23} Hg atoms, has a mass of 200.6 **grams**. The mole thus acts as the conversion from microscopic units (amu) to macroscopic units (grams).

The mole/mass relationship is also applicable to molecules. We simply add up the masses of the individual atoms in the formula, and one mole of that molecule has that sum of masses in grams. For instance, since there are two hydrogen atoms in a hydrogen molecule, a single H_2 molecule has a mass of $1.008 + 1.008 = 2.016$ amu, so a **mole** of hydrogen molecules has a mass of 2.016 grams. Note the difference: a mole of hydrogen **atoms** has a mass of 1.008 grams, but a mole of hydrogen **molecules** has a mass of 2.016 grams. It is important to start keeping track of the exact chemical identity of the material you are working with so you don't get confused. The following example shows how moles of atoms can be different from moles of molecules.

Example 5.1. How many moles of hydrogen atoms are there in (a) one mole of water, H_2O (b) 2.5 moles of benzene, C_6H_6 (c) one-half mole of hydrogen peroxide, H_2O_2.

Solutions. (a) Since the chemical formula for water shows that there are two atoms of hydrogen in every molecule of water, then in one mole of water there are a total of two moles of individual hydrogen atoms. (b) The formula for benzene indicates that there are six hydrogen atoms in every molecule of C_6H_6, so there are $6 \times 2.5 = 15$ moles of individual hydrogen atoms. (c) Like water, hydrogen peroxide has two hydrogen atoms in each molecule, so in one-half mole of H_2O_2 there are $2 \times \dfrac{1}{2} = 1$ mole of hydrogen atoms.

The mass of one mole of atoms of any element is called the atomic weight of that element. Atomic weights are typically listed in periodic tables. The mass of one mole of molecules, or one formula unit of an ionic compound, is called the molecular weight of that compound.* (Formula weights are sometimes used for ionic compounds.) We therefore have the relationship that

1 mole of material = 1 atomic or molecular weight of that material

We can use this equation to make a conversion factor between amount in moles and mass in grams. The next example shows how we do this.

Example 5.2. What is the mass of (a) 0.0550 moles of mercury, Hg (b) 2.75 moles of NaCl.

Solutions. (a) First, let us set up our conversion factors. Using the mole unit and its relationship to the atomic weight of an element:

$$1 \text{ mol Hg} = 200.6 \text{ g Hg}$$

---

* Although these terms use the word 'weight', they are actually masses. Weight is a force due to gravity; mass is an inherent property of matter. However, the terms are so embedded in science that it would be fruitless to try to change them.

Notice that we are being explicit about "moles of what" and "grams of what." It is important to keep track of this. To construct a conversion factor from the above equation, remember that we are given an

amount in moles at the beginning, and that we want to cancel this out and convert to units of grams. We therefore make our conversion factor with the mole unit in the denominator:

$$\frac{200.6 \text{ g Hg}}{1 \text{ mol Hg}} = 1$$

This fraction **does** equal one, because one mole of mercury **is** 200.6 grams of mercury! That is the atomic weight of mercury. We use this conversion factor algebraically just like any other conversion factor. Starting out with what we are given, we now convert from moles to grams:

$$0.0550 \text{ mol Hg} \cdot \frac{200.6 \text{ g Hg}}{\text{mol Hg}} = 11.0 \text{ g Hg}$$

Notice that the "mol Hg" cancels algebraically, like any other unit. Our final answer is limited to three significant figures. (b) In this case, we are dealing with a compound, so we need to determine the molecular weight of NaCl. Using the periodic table, we find that the atomic weight of sodium is 22.9898 and the atomic weight of chlorine is 35.453. Adding them together, we get 58.443 as the molecular weight for NaCl, which means that

$$1 \text{ mol NaCl} = 58.443 \text{ g NaCl}$$

Again, we want a conversion factor that allows us to cancel the "mol NaCl" unit in what we are given, so we use the above equation to construct the conversion factor

$$\frac{58.443 \text{ g NaCl}}{1 \text{ mol NaCl}}$$

Using this as a conversion factor, we get

$$2.75 \text{ mol NaCl} \cdot \frac{58.443 \text{ g NaCl}}{\text{mol NaCl}} = 161 \text{ g NaCl}$$

The previous example shows two things. First, we can construct conversion factors to go from units of moles to units of grams, using atomic or molecular weights. Part (a) used an atomic weight, and part (b) used a molecular weight, but the conversion itself was similar in both cases. Second, we must be extremely careful to keep track of what material we are working with. For instance, in part (a) above we found that one mole was equal to 200.6 grams, but in part (b) we found that one mole was equal to 58.443 grams. **That's because we were working with different materials!** Every chemical has its own characteristic atomic or molecular weight, and it is absolutely essential that you keep track of what material you are referring to when you perform these types of calculations. That's why we included the "Hg" and "NaCl" when we used "mole" or "gram" units: it helps us keep track of what material we are talking about. In the next few sections, the need to keep track of what material is being referred to will become even more obvious.

What is the relationship between moles and balanced chemical reactions? **A balanced chemical reaction can be considered as written in terms of moles of reactants and products, not molecules.** Therefore,

$$2\,H_2 \,+\, O_2 \,\rightarrow\, 2\,H_2O$$

can be spoken as "2 moles of hydrogen and 1 mole of oxygen react to make 2 moles of water." All balanced chemical reactions can be thought of as being balanced on a **molar** scale, not just an atomic or molecular scale.

Occasionally, one sees a chemical reaction that is balanced using fractional coefficients. This is not considered absolutely wrong, but some textbooks do not use them as a matter of course. For instance,

$$H_2 \,+\, \frac{1}{2}\,O_2 \,\rightarrow\, H_2O$$

is considered balanced. But how can we speak of half of a molecule? We don't: we speak of one-half of a **mole** of diatomic oxygen, which equals one mole of oxygen atoms. Since there is one mole of oxygen atoms in the products, the chemical reaction is balanced. **You should check with your textbook and your instructor to find out if using fractional coefficients is acceptable in your course.**

Mole-Mole Problems

A balanced chemical reaction is a source of various conversion factors that we can use to make a variety of calculations. Again, consider the following reaction:

$$2 H_2 + O_2 \rightarrow 2 H_2O$$

If we ask ourselves, "How many moles of oxygen will react with 2 moles of hydrogen," the answer is obvious. One mole of oxygen reacts with 2 moles of hydrogen. Suppose we are going to react 20 moles of hydrogen? We will react 10 moles of oxygen. We are using the 2:1 ratio given by the coefficients in the balanced chemical reaction.

One way of looking at this is that **2 moles of hydrogen are chemically equivalent to one mole of oxygen.** That is, we can write

$$2 \text{ mol } H_2 = 1 \text{ mol } O_2$$

as far as the above reaction is concerned. (This seems an unusual "equation" to write, but the **balanced** chemical reaction says that hydrogen and oxygen react in this proportion. We think of the "equivalence" as a chemical one that we can take advantage of mathematically.) This allows us to construct the following conversion factors:

$$\frac{2 \text{ mol } H_2}{1 \text{ mol } O_2} = \frac{1 \text{ mol } O_2}{1 \text{ mol } H_2} = 1$$

We can use these conversion factors to calculate the moles of one substance that will react with a given number of moles of another substance.

Example 5.3. How many moles of O_2 will react with 7.33 moles of H_2 to make water?

Solution. We will start our calculation with what we are given: 7.33 moles of H_2. We can then use the proper conversion factor, the one with "mol H_2" in the denominator, to convert to moles of O_2:

$$7.34 \text{ mol } H_2 \cdot \frac{1 \text{ mol } O_2}{2 \text{ mol } H_2} = 3.67 \text{ mol } O_2$$

Notice how the "mol H_2" unit cancels in the example above. Suppose, however, the conversion factor was written simply

$$\frac{1 \text{ mol}}{2 \text{ mol}}$$

There might be a temptation to cancel the "mol" units within this conversion factor. Conceptually, this would be incorrect: the "mol" unit in the numerator refers to moles of **oxygen**, whereas the "mol" unit in the denominator refers to moles of **hydrogen**. It is important that you are aware of this difference so you do not get your units confused. That's why we write "mol H_2" and "mol O_2," and you are encouraged to identify your units properly to minimize making errors.

In the balanced chemical reaction, **all** of the products and reactants are chemically equivalent, so there are many possible conversion factors you can construct. For the reaction of hydrogen and oxygen making water, we have

$$2 \text{ mol } H_2 = 1 \text{ mol } O_2 = 2 \text{ mol } H_2O$$

so we can devise conversions not only between reactants but also between reactants and products.

Example 5.4. How many moles of water are produced when 4.89 moles of hydrogen gas reacts with oxygen?

Solution. Since we are dealing with hydrogen and water, we need a conversion factor that involves these two chemicals. From the balanced chemical reaction, we see that two moles of hydrogen react to make two moles of water. Since we are given an amount of hydrogen and want an amount of water, we put moles of hydrogen in the denominator of the conversion factor and moles of water in the numerator:

$$\frac{2 \text{ mol } H_2O}{2 \text{ mol } H_2}$$

Solving the problem, we start with what we are given:

$$4.89 \text{ mol } H_2 \cdot \frac{2 \text{ mol } H_2O}{2 \text{ mol } H_2} = 4.89 \text{ mol } H_2O$$

These types of problems are called mole-mole problems, because they require you to calculate moles of one chemical from the given moles of another chemical in the balanced chemical reaction.

You must have a properly-balanced chemical reaction in order to work these types of problems! Furthermore, it must be the **right** reaction. Hydrogen and oxygen can also react to make hydrogen peroxide, H_2O_2:

$$H_2 + O_2 \rightarrow H_2O_2$$

where the reaction stoichiometry is different. Although this reaction is balanced, it is not the correct balanced reaction for the above examples because it does not have water as the product.

Calculations using complicated chemical reactions can be performed just as easily, as the next example shows.

Example 5.5. How many moles of manganese sulfate, $MnSO_4$, will be made as product if 2.89 moles of hydrogen peroxide, H_2O_2, were decomposed according to the following balanced chemical reaction:

$$2\, KMnO_4 + 5\, H_2O_2 + 3\, H_2SO_4 \rightarrow 2\, MnSO_4 + K_2SO_4 + 5\, O_2 + 8\, H_2O$$

Solution. While this is a more complicated chemical reaction, it is balanced, and we can construct proper conversion factors. The chemical reaction says that when 5 moles of H_2O_2 are reacted, 2 moles of $MnSO_4$ are produced. We therefore make the following conversion factor:

$$\frac{2 \text{ mol } MnSO_4}{5 \text{ mol } H_2O_2}$$

and we can calculate the answer:

$$2.89 \text{ mol } H_2O_2 \cdot \frac{2 \text{ mol } MnSO_4}{5 \text{ mol } H_2O_2} = 1.16 \text{ mol } MnSO_4$$

You should note that the numbers in the mole-to-mole conversion factors are considered exact numbers and are not considered when determining the number of significant figures in the final answer.

Mass-Mass Problems

It is a simple step to expand these problems to calculate masses of chemicals. Using the atomic or molecular weights, we can calculate the mass of a product or a reactant if we know its number of moles, and vice versa. These types of problems are called mass-mass problems. What we see, though, is that the mole unit still has a central place even in these kinds of problems.

Given the following balanced chemical reaction,

$$Mg + 2\,HCl \rightarrow MgCl_2 + H_2$$

suppose we want to calculate how many grams of hydrogen gas we can make from 100.0 grams of Mg metal. Because the balanced chemical reaction can be thought of in terms of moles, **not grams**, our first step is to calculate the number of moles of Mg we have. Using the fact that the atomic weight of Mg is 24.3 grams per mole to three significant figures, we have

$$100.0\ \cancel{g\,Mg} \cdot \frac{1\ \text{mol Mg}}{24.3\ \cancel{g\,Mg}} = 4.12\ \text{mol Mg}$$

where we have started the calculation with what we were given (100.0 g Mg) and have written our conversion factor so that the "g Mg" unit cancels.

Now we can use the balanced chemical reaction to determine how many moles of hydrogen gas is made. For every mole of Mg, one mole of H_2 is given off. Therefore, 1 mol Mg = 1 mol H_2 and we can convert to moles of hydrogen:

$$4.12\ \cancel{\text{mol Mg}} \cdot \frac{1\ \text{mol H}_2}{1\ \cancel{\text{mol Mg}}} = 4.12\ \text{mol H}_2$$

Note how the "mol Mg" units cancel, leaving "mol H_2" as the unit. The question asks for number of grams of hydrogen, so we must perform a third step to convert moles of hydrogen to grams of hydrogen. By adding up the atomic weights of the two hydrogen atoms, we find that one mole of hydrogen has a mass of 2.016 grams. We want to cancel the "mol H_2" unit, so that part of the conversion factor goes into the denominator:

$$4.12\ \cancel{\text{mol H}_2} \cdot \frac{2.016\ \text{g H}_2}{1\ \cancel{\text{mol H}_2}} = 8.31\ \text{g H}_2$$

So, we can calculate that we get a little over 8 grams of hydrogen gas when we react 100.0 grams of magnesium.

Problems like these can be done in many short steps, like we did above, or in one big step:

$$100.0 \text{ g Mg} \cdot \frac{1 \text{ mol Mg}}{24.3 \text{ g Mg}} \cdot \frac{1 \text{ mol H}_2}{1 \text{ mol Mg}} \cdot \frac{2.016 \text{ g H}_2}{1 \text{ mol H}_2} = 8.30 \text{ g H}_2$$

Notice the existence of truncation error in this problem, too. The answer should be about the same no matter which way you work out the problem. Practice with your calculator to make sure that you get 8.30 as your final answer. You should also check the expressions to see if the appropriate units all cancel correctly, if you work out a problem like this in one long calculation.

Other Types of Problems

The mole-mole and mass-mass problems are a large part of the problems you can work in chemistry. In a sense, they are all the same: you take an initial amount and convert it, using definitions and the balanced chemical equation.

These types of problems can be expanded to include other kinds of conversion factors, like density or molarity or the definition of mole. In fact, in chemistry there will be many equivalencies that can be used as conversion factors in problems such as these. The following examples illustrate some of them. You should be able to do any problem of this type, if given the appropriate information to use as conversion factors.

Example 5.6. How many milliliters of a 0.2515 M solution of potassium hydroxide, KOH, are required to react with 36.22 mL of 0.1889 M sulfuric acid, H_2SO_4? The balanced chemical reaction is

$$2 \text{ KOH} + H_2SO_4 \rightarrow 2 \text{ H}_2O + K_2SO_4$$

and occurs in aqueous solution.

Solution. Since we ultimately relate everything in terms of moles, we will start by using the molarity unit as a conversion from volume to moles for sulfuric acid. First, we convert the units of volume to liters:

$$36.22 \text{ mL} \cdot \frac{1 \text{ L}}{1000 \text{ mL}} = 0.036\ 22 \text{ L}$$

Now we use the definition of molarity to determine the number of moles of H_2SO_4 that are present:

$$0.1889 \text{ M} = \frac{\text{\# moles } H_2SO_4}{0.036\ 22 \text{ L } H_2SO_4}$$

$$\text{\# moles } H_2SO_4 = (0.1889 \text{ M}) \cdot (0.036\ 22 \text{ L})$$

$$\text{\# moles } H_2SO_4 = 0.006\ 842 \text{ mol}$$

Recall that since molarity is defined as $M = \dfrac{\text{mol}}{\text{L}}$, the product "M·L" equals "mol." Now we convert this number of moles of H_2SO_4 to the number of moles of KOH that will react with it. Using the fact that the balanced chemical reaction states that 2 moles of KOH are needed to react with each mole of H_2SO_4:

$$0.006\ 842 \text{ mol } H_2SO_4 \cdot \frac{2 \text{ mol KOH}}{1 \text{ mol } H_2SO_4} = 0.013\ 68 \text{ mol KOH}$$

Finally, knowing the molarity of the KOH solution, we can determine the number of mL needed to supply this number of moles of KOH. Again, using the definition of molarity:

$$0.2515 \text{ M KOH} = \frac{0.013\ 68 \text{ mol KOH}}{\text{\# L KOH}}$$

Solving for volume, we cross-multiply and get

$$\text{\# L KOH} = 0.054\ 39 \text{ L} = 54.39 \text{ mL}$$

where we have converted the final answer to milliliters implicitly.

Example 5.7. The density of dry air is approximately $1.057 \frac{g}{L}$ at 0°C and 1 atm pressure. Assuming that the average molecular weight of air is 28.8 grams per mole, estimate the number of gas molecules per cubic centimeter in air. Recall that there are 6.02×10^{23} molecules in a mole of molecules.

Solution. What we need is a conversion factor based on the following relationship:

$$1 \text{ mole air} = 28.8 \text{ grams of air}$$

We will also use the fact that $1 \text{ L} = 1000 \text{ cm}^3$. Starting with what we are given, the density, we first convert into units of moles per liter:

$$1.057 \frac{g}{L} \cdot \frac{1 \text{ mol}}{28.8 \, g} = 0.0367 \frac{\text{mol}}{L}$$

Now we convert to molecules per liter:

$$0.0367 \frac{\text{mol}}{L} \cdot \frac{6.02 \times 10^{23} \text{ molecules}}{1 \text{ mol}} = 2.21 \times 10^{22} \frac{\text{molecules}}{L}$$

Finally, we convert the liters to cubic centimeters, getting

$$2.21 \times 10^{22} \frac{\text{molecules}}{L} \cdot \frac{1 \, L}{1000 \text{ cm}^3} = 2.21 \times 10^{19} \frac{\text{molecules}}{\text{cm}^3}$$

as our final answer. We could have performed all conversions in one long line:

$$1.057 \frac{g}{L} \cdot \frac{1 \text{ mol}}{28.8 \text{ L}} \cdot \frac{6.02 \times 10^{23} \text{ molecules}}{1 \text{ mol}} \cdot \frac{1 \text{ L}}{1000 \text{ cm}^3} = 2.21 \times 10^{19} \frac{\text{molecules}}{\text{cm}^3}$$

The numerical answer would be the same. You should convince yourself that the units do cancel out properly!

Student Exercises

5.1. Balance the following reactions.

(a) $NaOH + H_2SO_4 \rightarrow Na_2SO_4 + H_2O$

(b) $BaCl_2 + Al_2(SO_4)_3 \rightarrow BaSO_4 + AlCl_3$

(c) $NaHCO_3 + HCl \rightarrow NaCl + H_2O + CO_2$

5.2. (a) How many moles of O atoms are there in 1.5 moles of glucose, $C_6H_{12}O_6$? (b) How many moles of H atoms are there in 0.50 moles of sucrose, $C_{12}H_{22}O_{11}$? (c) How many moles of total atoms are there in 2.0 moles of ethanol, C_2H_5OH?

5.3. (a) How many grams are there in 2.50 moles of NaOH? (b) How many moles are there in 100.0 grams of CO_2? (c) How many grams are there in 5.22×10^2 L of argon gas if each mole occupies 24.4 liters of volume? (d) How many moles are present in 25.9 mL of hexane, C_6H_{14}, if hexane has a density of 0.6603 $\frac{g}{mL}$?

5.4. Which contains more total atoms, one mole of $Al_2(CO_3)_3$ or two moles of $BaSO_4$?

5.5. Which contains more total atoms, 1.00×10^2 grams of NaCl or 1.00×10^2 grams of LiBr?

5.6. How many moles of water are given off by the metabolism of 1 mole of sucrose? The unbalanced chemical reaction is

$$C_{12}H_{22}O_{11} + O_2 \rightarrow CO_2 + H_2O$$

5.7. How many moles of carbon dioxide are given off by 5.00 moles of sodium bicarbonate which, when heated, decomposes according to the following balanced chemical reaction (which is why $NaHCO_3$ is sometimes used as a fire extinguisher):

$$2\ NaHCO_3 + heat \rightarrow Na_2CO_3 + H_2O + CO_2$$

5.8. If each mole of gas has a volume of 22.4 L at standard temperature and pressure, what volume of CO_2 is given off by the decomposition of 1.00×10^2 g of sodium bicarbonate? See the above problem for the balanced chemical reaction.

5.9. Ammonia is produced industrially by the following gas-phase reaction:

$$N_2 + 3\ H_2 \rightarrow 2\ NH_3$$

(a) How many grams of ammonia will be produced by the complete reaction of 150. (1.50×10^2) grams of hydrogen gas? (b) How many grams of nitrogen are necessary to react with 150. (1.50×10^2) grams of hydrogen? (c) Do these masses satisfy the Law of Conservation of Mass?

5.10. How many milliliters of 0.1107 M HCl solution are needed to react with 1.022 grams of $Ba(OH)_2$? The unbalanced chemical reaction is

$$HCl + Ba(OH)_2 \rightarrow BaCl_2 + H_2O$$

5.11. If the density of 1.450 M HCl solution is 1.065 $\frac{g}{mL}$, what mass of HCl solution is needed to react with 45.02 mL of 2.070 M $Ca(OH)_2$? The balanced chemical reaction is

$$2 HCl + Ca(OH)_2 \rightarrow CaCl_2 + 2 H_2O$$

5.12. The atmosphere on the planet Mars consists almost solely of CO_2 and has approximately 6.07×10^{16} molecules per cubic centimeter. Calculate the density of Mars' atmosphere.

Answers to Student Exercises

5.1. (a) $2 \, NaOH + H_2SO_4 \rightarrow Na_2SO_4 + 2 \, H_2O$

(b) $3 \, BaCl_2 + Al_2(SO_4)_3 \rightarrow 3 \, BaSO_4 + 2 \, AlCl_3$

(c) $NaHCO_3 + HCl \rightarrow NaCl + H_2O + CO_2$ (The reaction is already balanced.)

5.2. (a) There are 9.0 moles of O atoms. (b) There are 11 moles of H atoms. (c) There are 18 moles of total atoms.

5.3. (a) There are 100. (1.00×10^2) grams of NaOH. (b) There are 2.273 moles of CO_2. (c) There are 855 grams of Ar gas. (d) There are 0.198 moles of hexane.

5.4. One mole of $Al_2(CO_3)_3$ contains more moles of atoms (14) than do 2 moles of $BaSO_4$ (12).

5.5. 100. grams of NaCl contains 2.06×10^{24} atoms, while 100. grams of LiBr contains 1.39×10^{24} atoms. The 100. grams of NaCl has more total atoms.

5.6. You should get 11 moles of water for every mole of sucrose metabolized.

5.7. You will get 2.50 moles of CO_2 for every 5.00 moles of $NaHCO_3$ that decomposes.

5.8. 13.3 L of carbon dioxide.

5.9. (a) 844 grams of ammonia (b) 694 grams of N_2. (c) Yes, because the total mass of reactants (694 + 150) equals the total mass of products (844).

5.10. We will need 107.8 mL of HCl solution.

5.11. We will need 136.9 grams of HCl solution.

5.12. The density of the Martian atmosphere is 4.44×10^{-3} grams per cubic centimeter, or 0.004 44 $\frac{g}{cm^3}$. This compares to a density of about 1.05 $\frac{g}{cm^3}$ for Earth's atmosphere!

Chapter 6. Using Mathematical Formulas

Introduction

Another important type of math problem that you will need to be able to do in working chemistry problems is one in which a specific equation or formula must be used. Many of these formulas have names: Boyle's law, Charles' law, the ideal gas law, the Nernst equation. Many of them don't have names. Most of them will probably have to be memorized. But even if the equations are known (whether they are given to you or if you know them by memory), you still need to put quantities into the equation properly and to evaluate the unknown quantity correctly. Although we reviewed some of the algebra necessary to solve for unknowns in equations in Chapter 2, in this chapter we will focus more on formulas and how to use them.

Formulas

In the mathematical sense, a <u>formula</u> is an equation that relates different quantities to each other. (Chemistry also uses <u>chemical</u> <u>formulas</u> to indicate the compositions of elements and compounds, but we are not using the word "formula" in that way here.) Many of the definitions we have used in previous chapters are in fact very simple formulas. The definition of molarity,

$$\text{molarity} = \frac{\text{moles of solute}}{\text{liters of solution}}$$

can be considered a formula. It relates a concentration, in units of molarity, to the number of moles of solute and the volume of the solution in liters.

Most of the mathematical formulas we use in chemistry either relate seemingly non-related quantities to each other or allow us to determine how a quantity changes with other changes in conditions. For example, the ideal gas law is

$$PV = nRT$$

and relates the pressure (P), volume (V), number of moles (n), and the temperature (T) of a sample of gas. R is called the <u>ideal</u> <u>gas</u> <u>law</u> <u>constant</u> and is the proportionality constant that relates all of these quantities of the gas. Most formulas are expressed in terms of variables that represent various quantities and make the equation easier to write. The ideal gas law could be written as "(pressure)(volume) = (number of moles)(ideal gas law constant)(temperature)", but "$PV = nRT$" is more compact. In this case, the formula relates **instantaneous** conditions, or conditions of the gas as it exists right now. As an example of a formula that relates a change in conditions, consider Boyle's law,

$$P_1 V_1 = P_2 V_2$$

which relates an initial pressure and volume P_1 and V_1 with the values of some final pressure and volume P_2 and V_2, which have both changed in value. (Boyle's law presumes that the both temperature and amount of the gas stay constant as the pressure and volume change.) Formulas like this relate conditions of a system that change and show that some of the measurable quantities that describe a system change together, in a certain relationship.

Most formulas simply require that we substitute values for each of the variables in the equation except for one, and then algebraically solve for that one unknown variable. "That's easy," many students think. But many students don't do it **properly**. They get the wrong answer, and don't understand why. Although we will not be able to cover every possible formula you might encounter, hopefully the ideas that this chapter covers will help you properly use any formula to determine the correct desired quantity.

Plugging In Properly

In virtually all cases when working with a formula, you will know – or can find – the value for every variable in the equation except for one. This is usually the variable you will be asked to determine. What you need to be able to do is to properly substitute values for the variables, then perform some algebra to solve for the unknown quantity.

The first step to being able to use a formula properly is so basic that most people don't even recognize its importance: **know what the variables stand for.** In the ideal gas law equation given above, we defined what P, V, n, R, and T were. When you know what the variables stand for, you can substitute the proper numbers and units in the correct places in the formula. If you do not know what the variables stand for, there would be no way you could properly use the formula to determine some unknown quantity.

Why is this important to understand? Because science deals with a lot of different quantities, each of which has its own specific variable. Some of these variables may not be obvious, either. Using "P" for pressure and "V" for volume may be obvious, but using "R" for the ideal gas law constant is not. (The letter "r" doesn't even appear in the phrase "ideal gas law constant"!) Furthermore, because of the vast number of quantities that science defines, we are forced to use the same variable for different quantities. *R*, for example, is used to symbolize the ideal gas law constant, the Rydberg constant (used to discuss electronic structure of atoms), and roentgens (a unit of radiation exposure). **It is crucial that you know what the variables stand for in any formula you use.**

When you replace a variable by a specific quantity, you do not just substitute the number from the quantity. **You must substitute the unit of that quantity also.** We will see shortly how important units are in working with formulas.

Example 6.1. Given that a 0.773 mol sample of a gas has a temperature of 298 K, a volume of 12.7 L, and a pressure of 1.49 atm, rewrite the ideal gas law in terms of these values. The constant *R* has a value of $0.082\,05\ \dfrac{\text{L} \cdot \text{atm}}{\text{mol} \cdot \text{K}}$.

Solution. The pressure and volume are multiplied together on one side of the equation, whereas the amount, temperature, and *R* are multiplied together on the other side of the equation. We get

$$\underbrace{(1.49\ \text{atm})}_{P}\underbrace{(12.7\ \text{L})}_{V} \stackrel{?}{=} \underbrace{(0.773\ \text{mol})}_{n}\overbrace{\left(0.08205\ \frac{\text{L} \cdot \text{atm}}{\text{mol} \cdot \text{K}}\right)}^{R}\underbrace{(298\ \text{K})}_{T}$$

where the parentheses show the correct values substituted for the variables. Keep in mind that because multiplication is commutative, you don't have to write the values in the exact order as the variables are in the formula. You just have to multiply pressure and volume on one side and equate that product to the amount, the temperature, and the *R* values.

While the above example may have seemed trivial, there is another point to be made. When substituting into formulas, **the units must be consistent in all parts of the formula**. Let us take a closer look at the expression above. Units will cancel algebraically in two different ways: first, if the same unit is in both the numerator and denominator on the same side of an equation, and second, if the same unit were on both sides in the numerator or on both sides in the denominator of the equation.

Let us see how the units work in the formula in Example 6.1. On the right side, the units for temperature (degrees Kelvin) and amount (moles) cancel because they are in the numerator and denominator on the same side of the equation:

$$(1.49\ \text{atm})(12.7\ \text{L}) = (0.773\ \cancel{\text{mol}})\left(0.082\,05\ \frac{\text{L atm}}{\cancel{\text{mol}}\,\cancel{K}}\right)(298\ \cancel{K})$$

Next, we notice that the units "liters" and "atmospheres" are in the numerator on both sides of the equation. Therefore, they cancel algebraically:

$$(1.49 \; \cancel{atm})(12.7 \; \cancel{L}) = (0.773)(0.082 \; 05 \; \cancel{L} \cdot \cancel{atm})(298)$$

(Here, we have omitted the mole and temperature units because we canceled them in the previous step.) What we have left are simply some numbers that we need to multiply together:

$$(1.49)(12.7) = (0.773)(0.082 \; 05)(298)$$

$$18.9 = 18.9$$

The final equation is what is called a "reflexive relationship" in algebra and is understood to be true (i.e. something equals itself). The point here is that the units of each type of quantity (volume, pressure, temperature, amount) are all consistent with each other. Units must be consistent when substituting quantities into a formula.

Let us consider an example where the units are **not** consistent with each other. Again, we will use the ideal gas law. If we are given a 0.001 90 mol sample of gas that has a volume of 55.8 mL, a pressure of 658 torr, and a temperature of 37.0°C, we **could** plug into the ideal gas law directly and get

$$(658 \text{ torr})(55.8 \text{ mL}) = (0.00190 \text{ mol})\left(0.082 \; 05 \frac{\text{L} \cdot \text{atm}}{\text{mol} \cdot \text{K}}\right)(37°\text{C})$$

The problem with doing this is that the units will not cancel! We are using two different units for pressure, two different units for volume, and two different units for temperature. Only the "mole" unit will cancel properly.

This is not a permanent problem, **since we can convert units** and make them consistent with each other. For most types of variables, it does not matter which unit we convert to as long as all variables of the same type (volume, pressure, energy, etc.) are expressed in the same units. The exception is temperature, which in most formulas should always be expressed in Kelvin units.

We can convert all pressure units to atmospheres (there are exactly 760 torr in 1 atm), and all volume units to liters, and get

$$(0.866 \text{ atm})(0.0558 \text{ L}) = (0.001\ 90 \text{ mol})\left(0.082\ 05 \frac{\text{L} \cdot \text{atm}}{\text{mol} \cdot \text{K}} \right)(310 \text{ K})$$

(where we have also converted temperature into Kelvin units). **Now** the formula is correctly set up, and the units cancel and the numbers combine to get

$$0.0483 = 0.0483$$

which we know is true.

Example 6.2. Instead of converting pressure and volume units to atm and L, convert all pressure and volume units to torr and mL and show that the formula is still an equality.

Solution. Only the value for R has units that need to be converted. We use the relationship between L and mL, and the fact that 760 torr = 1 atm:

$$0.082\ 05 \frac{\cancel{\text{L}} \cdot \cancel{\text{atm}}}{\text{mol} \cdot \text{K}} \cdot \frac{760 \text{ torr}}{1 \cancel{\text{atm}}} \cdot \frac{1000 \text{ mL}}{1 \cancel{\text{L}}} = 62,360 \frac{\text{mL} \cdot \text{torr}}{\text{mol} \cdot \text{K}}$$

Using this value for R (it's the same R, just different units), we get

$$(658 \text{ torr})(55.8 \text{ mL}) = (0.001\ 90 \text{ mol})\left(62,360 \frac{\text{mL} \cdot \text{torr}}{\text{mol} \cdot \text{K}} \right)(310 \text{ K})$$

The units all cancel, and we are left with

$$(659)(55.8) = (0.001\ 90)(62,360)(310)$$

$$36,700 = 36,700$$

which is still an equality. The number is different, but that's only because we are using different units.

The point is that it doesn't matter which units you convert, as long as the units are the same. Otherwise, you will not be able to perform the correct algebra with the units.

Using Formulas to Solve for Unknowns

Most of the time, we use formulas to solve for some unknown quantity. Typically, we know all but one of the variables – either they're given in the problem or we can look up data in tables – and we need to calculate that unknown variable.

First, of course, you need to know what formula to use to solve a problem. Next, you need to plug in all the information you have. You will need to make sure that all of the units are consistent, and that they will cancel so that the only unit left is an appropriate unit for the quantity you are seeking. Finally, you evaluate all of the numbers to calculate your final answer. Keep in mind that you may have to algebraically manipulate the formula so that the variable you are looking for is by itself in the numerator.

A simple equation might be

$$\Delta G° = -nFE°$$

where $\Delta G°$ is the change in the Gibbs free energy of an oxidation-reduction ("redox") reaction, n is the number of moles of electrons transferred in the reaction, $E°$ is the voltage (symbol = V) of the redox reaction, and F is the <u>Faraday</u> <u>constant</u> and equals 96,500 coulombs (symbol = C) per mole of electrons. If we are given that the number of moles of electrons transferred is 2, and that the voltage of the redox reaction is +1.10 volts, then we simply plug in to get

$$\Delta G° = -(2 \text{ moles of electrons}) \times \left(96,500 \, \frac{C}{\text{mole of electrons}} \right) \times (1.10 \text{ V})$$

The "moles of electrons" units cancel and we multiply the numbers to get

$$\Delta G° = -212,000 \text{ C·V}$$

These units look funny until we learn that a coulomb-volt is equal to a joule, a unit of energy. Therefore, we write

$$\Delta G° = -212,000 \text{ J} = -212 \text{ kJ}$$

If we are given that the $\Delta G°$ for a three-electron reaction is +23.07 kJ and want to calculate the $E°$ for the process, we will need to do some algebraic rearranging of the same formula. The rearranging can be done before or after the values are plugged into the formula. If the algebra is done correctly, it should not matter which way you do it.

For example, we can take the formula

$$\Delta G° = -nFE°$$

and divide both sides by $-n$ and F to get

$$E° = -\frac{\Delta G°}{nF}$$

and now plug in our known values, converting our $\Delta G°$ into units of J:

$$E° = -\frac{23,070 \text{ J}}{(2 \text{ mol e}^-)\left(96,500 \ \dfrac{\text{C}}{\text{mol e}^-}\right)} = -0.120\frac{\text{J}}{\text{C}} = -0.120 \text{ V}$$

Or, you can substitute into the original equation, cancel the units, and multiply and divide the numbers appropriately to isolate $E°$ all by itself in the numerator:

$$23,070 \text{ J} = -(2 \text{ mol of electrons})\left(96,500 \ \frac{\text{C}}{\text{mol of electrons}}\right)(E°)$$

$$23,070 \text{ J} = -193,000 \text{ C} \cdot E°$$

$$E° = -\frac{23,070 \text{ J}}{193,000 \text{ C}} = -0.120\frac{\text{J}}{\text{C}} = -0.120 \text{ V}$$

Either way, we get the same answer, **as long as the algebra is performed properly.** If you are uncertain about your algebra ability, it might be easier to plug the numbers into the original equation first, then cancel units and do algebra with the numbers, rather than the variables.

Example 6.3. The combined gas law relates the pressure, volume, and absolute temperature of a gas and how they change with respect to each other. If P_1, V_1, and T_1 were the initial conditions of the gas and they changed to some final conditions P_2, V_2, and T_2, the combined gas law requires that

$$\frac{P_1 V_1}{T_1} = \frac{P_2 V_2}{T_2}$$

If the initial conditions were 1.50 L, 874 torr, and 295 K, and the final conditions were 2.95 L and 1.78 atm, what is the final temperature?

Solution. First, we should take the time to recognize what values are given to us. From the units, we can determine that the initial volume V_1 = 1.50 L, the initial pressure P_1 = 874 torr, and the initial temperature T_1 = 295 K. The final volume V_2 = 2.95 L, the final pressure P_2 = 1.78 atm, and the final temperature T_2 is the unknown. Notice that the pressures are given in two different units, and we must convert one of the values so that both pressure values have the same units. Let us convert 1.78 atm to units of torr. Since 760 torr = 1 atm:

$$1.78 \text{ atm} \cdot \frac{760 \text{ torr}}{1 \text{ atm}} = 1350 \text{ torr}$$

We will use 1350 torr as our final pressure. Note that we could have converted the torr unit to atmospheres, but either way will give us the same final answer if we do all of the algebra correctly.

There are several ways to approach the formula in this example. Perhaps the easiest will be to cross-multiply the two fractions,

$$\frac{P_1 V_1}{T_1} \diagdown \frac{P_2 V_2}{T_2}$$

$$T_2 P_1 V_1 = T_1 P_2 V_2$$

and then divide both sides by $P_1 V_1$:

$$\frac{T_2 \cancel{P_1 V_1}}{\cancel{P_1 V_1}} = \frac{T_1 P_2 V_2}{P_1 V_2}$$

This isolates T_2 in the numerator on one side of the equation, all by itself:

$$T_2 = \frac{T_1 P_2 V_2}{P_1 V_1}$$

Substituting for all of our known values, we get

$$T_2 = \frac{(295 \text{ K})(1350 \text{ torr})(2.95 \text{ L})}{(874 \text{ torr})(1.50 \text{ L})}$$

Notice that all of the units cancel except for K, a unit of temperature, which is what we're looking for. The final answer, after multiplying and dividing all of the numbers, is

$$T_2 = 896 \text{ K}$$

You should verify this by entering all of these numbers into your calculator to see if you can get the same answer.

Again, it is important to keep track of the units: the units of your final answer **must** be an appropriate unit for the quantity you are calculating. In the above example, we found units of degrees Kelvin, which is the proper unit for temperature. If you are solving for a pressure, the final unit should be atm, torr, mmHg, bar, or some other unit of pressure. If you are calculating an energy, the final unit had better be calories, joules, kilojoules, etc. These are appropriate units of energy. If you do not get an appropriate unit, **then there is probably something wrong with your calculation.** In time, you will find that keeping track of your units is a tool you can use to help you keep track of the progress of your calculation.

Formulas can get more complicated. Consider the following example:

$$\Delta G = \Delta G^\circ + RT \ln Q$$

which allows you to calculate the change in the Gibbs free energy at any condition, ΔG, from the ΔG at standard conditions, ΔG° (notice the superscript $^\circ$), and a correction given by $RT \ln Q$. The "ln" refers to the natural logarithm, which we will look at in the next chapter. Q is called the reaction quotient It is the "products-over-reactants" expression and depends on the specific chemical reaction of interest. (For additional information about reaction quotients, consult your chemistry textbook.) For instance, for the ionization reaction

$$CH_3COOH \rightleftharpoons CH_3COO^- + H^+$$

the expression for Q is

$$Q = \frac{[H^+][CH_3COO^-]}{[CH_3COOH]}$$

The variables in brackets refer to concentrations of each chemical species. (For a more detailed discussion of Q and how it relates to reactions, consult your chemistry textbook.) So, for that chemical reaction – and **only** for that chemical reaction – the complete formula to calculate ΔG is

$$\Delta G = \Delta G^\circ + RT \ln \frac{[H^+][CH_3COO^-]}{[CH_3COOH]}$$

Although this formula is a little more complicated than the ones we have discussed previously, the same ideas apply: you should be able to solve for any one of the variables in the formula – even the individual concentrations of the chemical products and reactants.

This equation for ΔG brings up another issue concerning units. As an energy change, ΔG has units of joules or joules per mole. This means that the second term on the right, $RT \ln \frac{[H^+][CH_3COO^-]}{[CH_3COOH]}$, also must have units of joules or joules per mole. A logarithm is a pure number and doesn't have units, but R and T do. We usually have used R as equal to $0.082\,05 \ \frac{L \cdot atm}{mol \cdot K}$, and with temperature having units of K, the combination of the two variables has the overall unit of $\frac{L \cdot atm}{mol}$. This is not a typical unit of energy, and **not** equal to $\frac{J}{mol}$. However, we have seen that we can express R with different units. It will have a different numerical value, but that's OK. There is a value for R that has joule units: $R = 8.314 \ \frac{J}{mol \cdot K}$. When we use this value for R, the units work out properly for calculating a change in energy. Again, we stress the point that the units of the variables you substitute into formulas must be consistent.

Example 6.4. Given the chemical reaction

$$N_2 + 3\,H_2 \rightleftharpoons 2\,NH_3$$

if you were given that $\ln Q = 2.568$ at a temperature of 450°C and $\Delta G = -1.26$ kJ/mol, calculate $\Delta G°$ for the reaction.

Solution. Using the formula

$$\Delta G = \Delta G° + RT \ln Q$$

we should recognize that the problem gives us ΔG, T, $\ln Q$, and we know a value for R. We therefore plug in for everything but one variable, which is the one we are looking for: $\Delta G°$.

$$-1.26 \text{ kJ/mol} = \Delta G° + \left(8.314\,\frac{J}{mol \cdot K}\right)(450 + 273 \text{ K})\,(2.568)$$

where we have converted our temperature into degrees Kelvin. The temperature units on the right side cancel and we have

$$-1.26 \text{ kJ/mol} = \Delta G° + 15{,}400 \text{ J/mol}$$

In order to complete the problem, we need to convert units so that they are the same for all terms. Either the J/mol can be converted to kJ/mol, or the kJ/mol can be converted to J/mol. Let us convert the units on the last term to kJ/mol:

$$-1.26 \text{ kJ/mol} = \Delta G° + 15.4 \text{ kJ/mol}$$

Only when the units are the same can the numbers be combined. We subtract 15.4 kJ/mol from both sides in order to isolate the quantity we are solving for:

$$-16.7 \text{ kJ/mol} = \Delta G°$$

as our final answer.

In all of the above examples, we have been using many of the topics of previous chapters. We have limited our final answer to the tenths' place, according to the rules of significant figures. Notice, too, that in working this example, we followed the proper order of operations that were discussed in Chapter 2. The proper math skills must already be developed if a formula is going to be used to determine the correct answer.

Student Exercises

In many of the following exercises, you will be given a formula and some of the quantities, and you will then be asked to solve for one of the variables. Watch your units!

6.1. Another gas law, called <u>Amontons' law,</u> is given by the mathematical formula

$$\frac{V_1}{T_1} = \frac{V_2}{T_2}$$

In words, identify the variables. (HINT: see Example 6.3.)

6.2. Write a formula that expresses the following statement mathematically: "The force of gravity between two masses is equal to a proportionality constant times the product of the two masses and divided by the square of the distance between the masses."

6.3. Rewrite the ideal gas law to solve for R, the ideal gas law constant.

6.4. Use the equation $c = \lambda \cdot v$, where c is the speed of light (which is constant and equal to 2.9979×10^8 $\frac{m}{sec}$), λ is the wavelength of the light in units of meters, and v is the frequency in units of $\frac{1}{sec}$, to calculate the frequency of light that has a wavelength of 5.50×10^{-7} m. Do you see how the units work out?

6.5. Use the equation $E = h \cdot v$, where E is the energy of a single photon, v is the frequency in units of $\frac{1}{sec}$, and h is <u>Planck's</u> <u>constant</u> and is equal to 6.626×10^{-34} J·sec (joule-seconds), to calculate the energy of a single photon that has a wavelength of 5.50×10^{-7} m. See Exercise 6.4 for how to calculate the frequency v of the photon.

6.6. A <u>spectrum</u> is the absorption or emission of light of specific wavelengths by chemicals. The spectrum of hydrogen atoms is particularly simple because hydrogen is the simplest element. In fact, in 1885 Balmer showed that the wavelengths of visible light in the hydrogen spectrum could be predicted by the simple formula

$$\frac{1}{\lambda} = 10,973,700\left(\frac{1}{4} - \frac{1}{n^2}\right)\frac{1}{m}$$

where the unit for $\frac{1}{\lambda}$ is $\frac{1}{m}$. The variable n is an integer greater than 2: $n = 3, 4, 5$, etc. Calculate (to three significant figures) the wavelength λ of the line in the hydrogen spectrum corresponding to $n = 5$.

6.7. Use the combined gas law to determine the final pressure for a gas that has initial conditions 345 mL, 1.007 atm, and 25.0°C and final conditions 100°C and 1.28 L.

6.8. Use the ideal gas law to calculate the number of moles of gas that occupy a volume of 1.00 liters with a pressure of 1.0×10^3 torr and a temperature of 37.0°C. The additional information you need to solve this problem is located within the chapter text.

6.9. Use the equation $\Delta G° = -nFE°$ to calculate F for a redox reaction that involves two moles of electrons and has a voltage of 1.107 V and a $\Delta G°$ of -212.3 kJ.

6.10. The <u>Nernst</u> <u>equation</u> is a formula for calculating the voltage of a battery at non-standard conditions:

$$E = E° - \frac{RT}{nF} \ln Q$$

where E is the non-standard voltage, $E°$ is the voltage of the battery under standard conditions, and the rest of the variables have their normal meaning. (a) Identify the other variables R, T, n, F, and $\ln Q$. (b)

If the voltage of a battery is +0.453 V at a temperature of 37°C for a redox reaction that involves 5 moles of electrons and has $\ln Q = -3.775$, what is the voltage of the battery at standard conditions? (Note that the formula doesn't even require that you know what "standard conditions" are!) You will need to use $F = 96,500$ C/mol e⁻ and $R = 8.314 \dfrac{\text{J}}{\text{mol} \cdot \text{K}}$.

6.11. In 1913, scientist Neils Bohr derived a formula for the Rydberg constant, R, which is part of the formula that describes the hydrogen atom spectrum. (See Exercise 6.6 above.) Bohr found that

$$R = \frac{m_e e^4}{8\varepsilon_0^{\,2} h^2}$$

where the constants in the formula and their values are:

Symbol	Meaning	Value
m_e	mass of electron	9.109×10^{-31} kg
e	charge on electron	1.602×10^{-19} C
ε_0	permittivity of free space	$8.854 \times 10^{-12} \dfrac{\text{C}^2}{\text{J} \cdot \text{m}}$
h	Planck's constant	6.626×10^{-34} J·sec

(a) Substitute the values into the formula and calculate a value for R. You may have to be careful if your calculator does not allow three-digit powers of ten. What are the final units you get? (HINT: Remember that joule is a derived unit. What is its definition in terms of fundamental units?)

(b) The value of R in Exercise 6.6 is $10,973,700 \; \dfrac{1}{\text{m}}$. Using the equations $E = h\nu$ and $c = \lambda\nu$, can you convert your answer in (a) to this value and this unit?

Answers to Student Exercises

6.1. V_1 is the initial volume, T_1 is the initial temperature, V_2 is the final volume, and T_2 is the final temperature.

6.2. Your formula might look something like this: $F = k \cdot \dfrac{m_1 \cdot m_2}{r^2}$, where F is the force of gravity, k is the proportionality constant, m_1 and m_2 are the masses, and r is the distance between the masses. You might have used different symbols for the variables, but the form of your equation should be similar.

6.3. $R = \dfrac{PV}{nT}$

6.4. $v = 5.45 \times 10^{14} \dfrac{1}{\text{sec}}$, or 5.45×10^{14} sec^{-1}.

6.5. $E = 3.61 \times 10^{-19}$ J for a single photon.

6.6. $\lambda = 4.34 \times 10^{-7}$ m for the spectral line corresponding to $n = 5$. In case you're curious, that wavelength corresponds to light having a violet color, almost out of range of visible light.

6.7. The final pressure is 0.340 atm.

6.8. $n = 5.2 \times 10^{-2}$ mol, or 0.052 mol of gas.

6.9. Using the data given, we calculate a value for F as 95,890 C/mol e$^-$.

6.10. (a) R is the ideal gas law constant, T is the absolute temperature, n is the number of electrons transferred in the redox reaction, F is the Faraday constant, and ln Q is the natural logarithm of the reaction quotient, the normal "products-over-reactants" expression for a chemical reaction. (b) $E^\circ = 0.473$ V.

6.11. (a) Simply substituting into the expression, you should get 2.179×10^{-18} J. (b) Using the two equations, you should be able to get that $R = 10,970,000 \dfrac{1}{\text{m}}$, which to four significant figures is the same value for R that was used in the previous exercise.

Chapter 7. Advanced Math Topics

Introduction

Don't let the chapter title scare you! It's just that occasionally in general chemistry, there are some specialized math tools that must be used in order to work out a problem. Such topics include things like the quadratic formula, logarithms, exponentials, and roots. Since many of these topics are math skills that are only used in the latter part of a general chemistry sequence, we have put off discussing them until now.

Exponents

Most of the equations we have used so far have been composed of different variables multiplied, divided, added to, and subtracted from each other. Usually, the variable occurs once in the equation. Since it is understood that these variables have an exponent of one, we state that the variables are "raised to the first power." Variables that are raised to the first power are the most common variables we encounter, and one is the presumed exponent if no exponent is written explicitly.

When we multiply a variable by itself, like $V \times V$, we write it as V^2 and say that the variable is "raised to the second power" or that the variable is "squared." If three of the same variable are multiplied, like $V \times V \times V$, we write that as V^3, which is the variable "raised to the third power," or "cubed." Similarly, we have V^4 as "fourth power," V^5 as "fifth power," etc. (There are no common alternate names for variables raised to any power other than 2 or 3.)

If two different powers of the same variable are multiplied together, the result will be the variable raised to a power equal to the sum of the exponents. For example,

$$V^2 \cdot V^3 = V^{(2+3)} = V^5$$

This rule works with numbers as well as variables. The most common occurrence will be with powers of ten; for example,

$$10^5 \cdot 10^8 = 10^{(5+8)} = 10^{13}$$

For division of the same variable raised to different exponents, the result will be the variable raised to a power equal to the exponent of the numerator minus the exponent in the denominator. For example,

$$\frac{V^5}{V^2} = V^{(5-2)} = V^3$$

Negative numbers can also be used as exponents. A negative exponent implies the reciprocal of the variable raised to the positive exponent. For example,

$$V^{-1} \text{ means } \frac{1}{V^1} = \frac{1}{V}$$

In a similar fashion,

$$V^{-3} \text{ means } \frac{1}{V^3}$$

and so forth. If a variable with a negative exponent is in the denominator of a fraction, its reciprocal places it in the **numerator** with a positive exponent:

$$\frac{1}{V^{-3}} = \frac{1}{1/V^3} = 1 \div \frac{1}{V^3} = 1 \times \frac{V^3}{1} = V^3$$

In the next to last step, we have used the fact that division is equal to multiplication by the reciprocal. (See Chapter 2.)

Example 7.1. Simplify the following expression so that all variables have positive exponents.

$$\frac{x^{-3}y^2}{w^{-1}z^{-4}}$$

Solution. In order to see the individual changes better, we will separate the four variables in the expression so that every variable in the numerator is written as a fraction over one and the variables in the denominator are written as one divided by the variable.

$$\frac{x^{-3}y^2}{w^{-1}z^{-4}} = \frac{x^{-3}}{1} \cdot \frac{y^2}{1} \cdot \frac{1}{w^{-1}} \cdot \frac{1}{z^{-4}}$$

(You should verify that this is correct.) The negative exponents on x, w, and z mean that we can rewrite those variables as the reciprocal raised to the positive power. Rewriting:

$$\frac{x^{-3}y^2}{w^{-1}z^{-4}} = \frac{1}{x^3} \cdot \frac{y^2}{1} \cdot \frac{w^1}{1} \cdot \frac{z^4}{1}$$

Notice that the y^2 term did not change. Recombining the four terms into a single fraction, we get

$$\frac{y^2 w^1 z^4}{x^3}$$

as our final answer with all positive exponents. Normally, the "1" exponent on the w variable is not written explicitly, but it is included here to illustrate the principle.

Variables are not the only things that can have negative exponents; units can as well. This might seem unusual unless you remember that units can also be in denominators of fractions and can be rewritten in a numerator if given a negative exponent. In fact, we have already worked with a unit that can be expressed with a negative exponent. When we consider the frequency of a wave (of light, for example), we speak of the "number of waves per second" that pass a particular point. The "number of waves" is simply a number, but the "per second" is a unit. Since frequency has no unit in the numerator, only in the denominator, frequency has units of $\frac{1}{\sec}$. Using negative exponents, this unit is sometimes written as $\sec^{-1}$ or s^{-1}. (Science defines the unit "hertz" as equal to $\frac{1}{\sec}$. The hertz unit has the abbreviation "Hz" and is named after Heinrich Hertz, who discovered radio waves.)

Derived units can also be expressed using negative exponents. For example, the unit $\frac{kJ}{mol}$ can be rewritten as

$$kJ \cdot \frac{1}{mol} = kJ \cdot mol^{-1}$$

You might occasionally see "$kJ \cdot mol^{-1}$" instead of "$\frac{kJ}{mol}$," but you should recognize that the two expressions mean the same thing.

Example 7.2. Write the fundamental units for the newton so that all units are in the numerator, with the appropriate units having negative exponents.

Solution. Recall the definition of the newton:

$$N = \frac{kg \cdot m^2}{sec^2}$$

We will rewrite the fraction like we did in Example 7.1:

$$\frac{kg \cdot m^2}{sec^2} = \frac{kg}{1} \cdot \frac{m^2}{1} \cdot \frac{1}{sec^2}$$

$$= \frac{kg}{1} \cdot \frac{m^2}{1} \cdot \frac{sec^{-2}}{1}$$

$$= kg \cdot m^2 \cdot sec^{-2}$$

You may think that this is an extreme example. However, many books, articles, and reference materials express complicated units for quantities and constants in this manner. Can you find any examples in your own textbook?

---

http://owl.thomsonlearning.com: Ch 0-2c Math: Exponents

The Quadratic Formula

In certain problems dealing with equilibrium concentrations, an equation arises that has an unknown – usually called "x" – raised to the second power. There may or may not be another term in the equation with the variable x raised to the first power, and there may or may not be a constant as part of the equation. The equation is equal to zero. For example, one such equation may look like this:

$$x^2 - 2x + 1 = 0$$

Any equation that is the combination of terms where the highest exponent on the variable is 2 is called a quadratic equation. A quadratic equation can be thought of as a combination of three powers of x: x raised to the second power, x raised to the first power, and x raised to the zero power. (Recall that anything raised to the zero power is one.) A proper quadratic equation equals zero, which may require that an equation be algebraically rewritten. The numbers that precede each power of x are called coefficients. In the above quadratic equation, the coefficient on the x^2 term is understood to be 1. The coefficient on the x term (where here the exponent of one is understood) is –2, and the coefficient on the x^0 term is 1. The letters a, b, and c are often used to represent the x^2, x^1, and x^0 coefficients, so any general quadratic equation can be written as

$$ax^2 + bx + c = 0$$

Understand that a, b, and c can be positive, negative, zero, and are not necessarily integers. (However, it's sometimes easier to use the quadratic formula when the equation has integer coefficients, so it is common to rewrite equations so that they have all-integer coefficients.)

Example 7.3. Rewrite the following expression as a proper quadratic equation with integer coefficients, and identify a, b, and c.

$$x^2 - 4x = -3 - 2x + \frac{2}{5}x^2$$

Solution. If we want all integers for coefficients, we must algebraically eliminate the fraction in the last term. To do this, we multiply both sides of the equation by 5:

$$5\left(x^2 - 4x\right) = 5\left(-3 - 2x + \frac{2}{5}x^2\right)$$

$$5x^2 - 20x = -15 - 10x + 2x^2$$

In order to bring all terms to the left side of the equation, we subtract $2x^2$ from both sides, add $10x$ to both sides, and add 15 to both sides. We get

$$(5x^2 - 2x^2) + (-20x + 10x) + 15 = (-15 + 15) + (-10x + 10x) + (2x^2 - 2x^2)$$

$$3x^2 - 10x + 15 = 0$$

This is our proper quadratic formula with integer coefficients. This means that $a = 3$, $b = -10$, and $c = 15$. You should verify that this is correct.

Because any quadratic equation is an equation, it is expected that there is some number which, when substituted for x, makes the equation, in fact, equal zero. Consider the equation

$$x^2 - 2x + 1 = 0$$

If we substitute 1 for x, we have

$$1^2 - 2(1) + 1 = 1 - 2 + 1 = 0$$

That is, when 1 is substituted for x on the left side, evaluating each term and combining them **does** give zero as the final answer. The expression "$x = 1$" is called a <u>solution</u> to the quadratic equation. In chemistry, when we get quadratic equations in setting up some problems, our ultimate task is to find the solutions to the equation.

Solutions to quadratic equations can be almost any number. They can be positive or negative, integer or non-integer. They might even be <u>imaginary</u>, which is a number that involves the square root of -1. (The symbol i is used to symbolize $\sqrt{-1}$.) Imaginary numbers arise when you try to take the square root of negative numbers. For example, for the quadratic equation

$$x^2 + 1 = 0$$

where $a = 1$, $b = 0$, and $c = 1$, we solve by bringing the 1 to the other side of the equation:

$$x^2 = -1$$

Finally, we simply take the square root of both sides:

$$x = \sqrt{-1} \equiv i$$

The imaginary number i does not have physical significance, but it does have important mathematical significance.

In chemistry, however, we are dealing with things that do have physical significance, like amounts and pressures and concentrations of chemicals. Therefore, in chemistry, we limit allowable solutions to our quadratic equations as those solutions **that give physically measurable values**. This means that all of our final pressures and concentrations and amounts are positive (although depending on how we define our unknown x, its value may be negative), and **imaginary solutions are not allowed**. These restrictions simplify our attempts to find solutions using quadratic equations.

But how do we find solutions to quadratic equations? Not all of them will be solvable by just looking at the equation and plugging in simple numbers like 1 or 2 or –1. However, we have a useful tool: the quadratic formula. Any equation that has exponents on its variable has a maximum number of possible different solutions equal to the highest-numbered exponent. For quadratic equations, that means that there will be a maximum of two different solutions. (The word "different" is important, because both solutions may be the same. For the equation $x^2 - 2x + 1 = 0$, both of the solutions are $x = 1$.) For any general quadratic formula given by the general equation

$$ax^2 + bx + c = 0$$

the two roots x_1 and x_2 are given by the expressions

$$x_1 = \frac{-b + \sqrt{b^2 - 4ac}}{2a}$$

$$x_1 = \frac{-b - \sqrt{b^2 - 4ac}}{2a}$$

The "$\sqrt{}$" sign means "take the square root of the expression inside." In this case, the expression is "$b^2 - 4ac$." You should have a special key on your calculator for evaluating the square root of a number. (Roots are considered in a later section.) Keep in mind that a, b, and c might be negative numbers! Since the only difference in the two expressions is the + or – sign in front of the square root term in the numerator, the solutions are generalized using the $\pm$ symbol and are usually seen as

$$x = \frac{-b \pm \sqrt{b^2 - 4ac}}{2a}$$

Therefore, in order to determine the solutions to your quadratic equation, all you need to know are the coefficients of the equation, which has been written to equal zero.

Example 7.4. Show that $x = 1$ is the only unique solution to the quadratic equation $x^2 - 2x + 1 = 0$.

Solution. Since both solutions have $\sqrt{b^2 - 4ac}$ in them, we will evaluate that part first. Since $a = 1$, $b = -2$, and $c = 1$, we have

$$\sqrt{b^2 - 4ac} = \sqrt{(-2)^2 - (4 \cdot 1 \cdot 1)}$$

$$= \sqrt{4 - 4} = \sqrt{0} = 0$$

We therefore have

$$x_1 = \frac{-(-2) + 0}{2 \cdot 1}$$

$$x_2 = \frac{-(-2) - 0}{2 \cdot 1}$$

Solving:

$$x_1 = \frac{2}{2} = 1 \quad \text{and} \quad x_2 = \frac{2}{2} = 1$$

The only unique solution is simply $x = 1$, which agrees with the solution we presented earlier.

The numbers won't always be as simple as in Example 7.4. However, the procedure for determining the solutions for a quadratic equation are the same **no matter how complicated the numbers look**. Therefore, once you have mastered using the quadratic formula, you can solve any quadratic equation. In chemistry, you should never get an imaginary number for your answer. They will always be real numbers.

Example 7.5. In working a problem involving a chemical reaction at equilibrium, the following equation is determined:

$$\frac{x^2}{0.02 - x} = 1.8 \times 10^{-5}$$

What are the possible values for x?

Solution. First, we should rewrite this equation into the proper format for a quadratic equation. We multiply through by $0.02 - x$, and then bring all terms over to one side to get an equation that equals zero.

$$x^2 = (0.02 - x)(1.8 \times 10^{-5})$$

$$x^2 = 3.6 \times 10^{-7} - (1.8 \times 10^{-5})x$$

$$x^2 + (1.8 \times 10^{-5})x - 3.6 \times 10^{-7} = 0$$

This shows us that $a = 1$, $b = 1.8 \times 10^{-5}$, and $c = -3.6 \times 10^{-7}$. (Notice that c is negative!) We now substitute these numbers into the quadratic formula:

$$x = \frac{-1.8 \times 10^{-5} \pm \sqrt{(1.8 \times 10^{-5})^2 - 4(1)(-3.6 \times 10^{-7})}}{2 \cdot 1}$$

$$x = \frac{-1.8 \times 10^{-5} \pm \sqrt{(3.24 \times 10^{-10} + 1.44 \times 10^{-6})}}{2}$$

$$x = \frac{-1.8 \times 10^{-5} \pm 1.2 \times 10^{-3}}{2}$$

By using the + sign in the numerator, we get

$$x = \frac{1.182 \times 10^{-3}}{2} = 5.9 \times 10^{-4}$$

By using the − sign in the numerator, we get the second solution:

$$x = \frac{-1.218 \cdot 10^{-3}}{2} = -6.1 \cdot 10^{-4}$$

Therefore, our solutions are $x = 5.9 \times 10^{-4}$ and $x = -6.1 \times 10^{-4}$. If these x's are representing solution concentrations, we need to recognize that it is not possible to have a **negative** concentration, and so we should omit the second solution. However, considerations like that will depend on the exact nature of the problem you are doing. Both answers are limited to two significant figures, and if you substitute them into the original quadratic equation, you might find evidence of truncation error (i.e. the answer may differ from zero by a small amount).

Consult your textbook for specific chemistry problems that use the quadratic formula to solve for an unknown. You can find many of them in your text where chemical equilibrium is discussed.

http://owl.thomsonlearning.com: Ch 0-5 Math: Quadratic Equations

Neglecting Terms

Earlier, when we set up the expression

$$\frac{x^2}{0.02 - x} = 1.8 \times 10^{-5}$$

we solved for x exactly by multiplying through with the denominator and expressing the equation as a proper quadratic equation. Consider the denominator, however: $0.02 - x$. We are subtracting some unknown amount, x, from 0.02. If the true value of the unknown x is very small compared to 0.02, then we are not affecting the absolute value of 0.02 all that much; in fact, if x is very small compared to 0.02, we are not going to change the final answer much if we simply neglect the x in the denominator:

$$\frac{x^2}{0.02 - x} = 1.8 \times 10^{-5}$$

neglect x **if it is small compared to 0.02**

In doing this, the algebra becomes a little simpler, and the final answer does not change much from the earlier answer:

$$\frac{x^2}{0.02} = 1.8 \times 10^{-5}$$

$$x^2 = (0.02)(1.8 \times 10^{-5})$$

$$x^2 = 3.6 \times 10^{-7}$$

$$x = 6.0 \times 10^{-4}$$

This answer is very close to the answer we got when we used the quadratic formula, and it was much simpler to determine.

So: when can we neglect a term like this? There are two rules:

- **Terms can only be neglected when they are being added or subtracted to another term.** You cannot neglect a term that is multiplying other terms.
- **Terms can only be neglected if they are small with respect to the other terms they are being added to or subtracted from.** How do you know if such terms will be small with respect to the other terms? Technically, you don't. You should always check your final answer for x and compare it to the term that is supposedly much larger than x. A good rule of thumb is that if x is expected to be much less than 10% of the larger value, you can save time by neglecting it and still come up with a reasonable final answer. Your textbook or instructor may have other rules, so it is wise to check.

Of course, neglecting terms is an approximation that will ultimately change your final answer. But if the terms are quite different in magnitude, you can save yourself a lot of time by simplifying the mathematics. The following two examples show cases where neglecting the x works, and then doesn't work.

Example 7.6. Solve the expression for x by (a) using the quadratic formula, and (b) neglecting x and solving. Compare your answers.

$$\frac{x^2}{1-x} = 1 \times 10^{-10}$$

Solution. (a) We first rearrange the equation into a proper quadratic equation:

$$x^2 + 1 \times 10^{-10}x - 1 \times 10^{-10} = 0$$

Substituting into the quadratic formula, we find

$$x = \frac{-1 \times 10^{-10} \pm \sqrt{(1 \times 10^{-10})^2 - 4(1)(-1 \times 10^{-10})}}{2 \cdot 1}$$

Solving, we get

$$x = -1 \times 10^{-5} \text{ and } x = 1 \times 10^{-5}$$

(Actually, the answers are $x = -0.000\ 010\ 000\ 1$ and $+0.000\ 009\ 999\ 9$, but we are keeping our answers to one significant figure.)

(b) By neglecting x in the denominator, we get

$$\frac{x^2}{1} = 1 \times 10^{-10}$$

This is easily solvable as

$$x = 1 \times 10^{-5} \text{ and } x = -1 \times 10^{-5}$$

As you can see, neglecting the x in the denominator did not practically change our final answer. Keep in mind that you should check your final answer and compare it with the numbers in the problem to see if neglecting the x was justified!

Example 7.7. Solve the expression for x by (a) using the quadratic formula, and (b) neglecting x and solving. Compare your answers.

$$\frac{x^2}{1.00 - x} = 1.00 \times 10^{-1}$$

Solution. (a) By writing 1.00×10^{-1} as 0.100, we can save ourselves some work. Rewriting as a proper quadratic equation:

$$x^2 + 0.100x - 0.100 = 0$$

Using the quadratic equation:

$$x = \frac{-0.100 \pm \sqrt{(0.100)^2 - 4(1)(-0.100)}}{2 \cdot 1}$$

Solving, we find that

$$x = 0.270 \text{ and } x = -0.370$$

(b) Neglecting the x in the denominator, we get

$$\frac{x^2}{1.00} = 1.00 \times 10^{-1}$$

Again, this is relatively easy to solve; we simply take the square root of 0.100:

$$x = 0.316 \text{ and } x = -0.316$$

We are off by about 15% on one solution and 16% on the other. These answers would probably not even be considered correct on a quiz or exam, because they are so far off (especially when compared to the previous example!). In this case, our unknown x is not less than 10% of the number it is subtracted from in the original problem, and neglecting the x was not the appropriate thing to do.

Students always have the question, "Should I neglect the x?" Unfortunately, this is one question where a simple answer won't work, in spite of the 10 Percent Rule mentioned above. Theoretically, it is best to try to solve the complete quadratic equation. Practically, in time and with practice, you will develop the sophistication to know when to neglect a term and when to not. Ask your instructor or consult your textbook for specific chemistry problems that require such decisions.

Roots

Most people are familiar with square roots (in fact, we have already used them in problems). The concept of a square root is, "What number '*x*' times itself equals another number '*y*'?" The number *x* is said to be the <u>square</u> <u>root</u> of *y*. It is written as

$$x = \sqrt{y}$$

For example, the square root of 9 is 3:

$$3 = \sqrt{9}$$

We know this because we can square both sides of the equation and get an equality again:

$$3^2 = 9$$

Negative numbers can also be square roots. (However, the square root of negative numbers brings up imaginary numbers, which we will not consider here.) Since we understand that $(-3)^2$ also equals nine, then it is proper to think of -3 as the square root of 9 also:

$$-3 = \sqrt{9}$$

Technically, when taking a square root of a number, **the positive and negative square root must be considered.** In many cases involving physical quantities, the negative square root is not an acceptable answer, but you should keep in mind that mathematically square roots can be positive **and** negative. Many people already know the numbers that have integral square roots (i.e. 1, 4, 9, 16, 25, 36, etc.), but all numbers have square roots. They are just not integers, and usually are evaluated using a calculator. For example, the square root of 10 is 3.162 277 6......

We can also represent square roots using exponents. Since we designate the square of a number with a superscripted "2", we designate the square root of a number using the reciprocal of 2: $\frac{1}{2}$. Thus, $\frac{1}{2}$ is used as a superscript, or an exponent, to indicate a square root. For example,

$$10^{1/2} = \sqrt{10} = 3.162\ 277\ 6.....$$

The rules of positive and negative exponents hold for these fractional exponents, too.

Example 7.8. What is the value of $10^{-1/2}$?

Solution. From the rules of negative exponents, $10^{-1/2}$ equals

$$\frac{1}{10^{1/2}} = \frac{1}{\sqrt{10}} = \frac{1}{3.162\ 277...} = 0.316\ 277...$$

Can you evaluate this on your calculator properly?

A square root is based on a squared number, a number multiplied by itself. A square root is the inverse operation of a square, so that taking a number, squaring it, and then taking the square root of the resulting number regenerates the original number (and vice versa). So,

$$x = \left(\sqrt{x}\right)^2 = \sqrt{x^2} = x$$

A square and a square root cancel each other.

The same thing holds true for other, larger powers. They, too, have their inverse. For example, the inverse of a number raised to the third power, or cubed, is called a cube root and is indicated by the symbol $\sqrt[3]{\ }$. So, for example, if $3 \times 3 \times 3 = 27$, then the cube root of 27, or $\sqrt[3]{27}$, is 3. Again, most numbers have non-integer cube roots, and we usually evaluate them using a calculator. Other roots – fourth roots, fifth roots, etc. – are also possible but are rare in chemistry.

Cube roots can also be written as fractional exponents. Since we write a cube with a 3 as a superscript, we indicate a cube root as a $\frac{1}{3}$ as a superscript. For example, $x^{1/3}$ implies the cube root of x. Cube roots cancel an exponent of 3, just like square roots cancel an exponent of 2.

Many calculators have a square key as well as a square root key: $\mathbf{x^2}$ and $\sqrt{\ }$, usually. However, most calculator don't have specific keys for cube roots or larger roots. Some calculators have keys that look like $\sqrt[\square]{\square}$, where you have to enter the root (the box outside the sign) and the number you're taking the root of (under the sign). On other calculators, you have to use a key that looks like $\mathbf{y^x}$ and enter the original number and the root in fractional form (i.e. **0.5** for a square root, **0.333 333 333 3...** for a cube root, **0.25** for a fourth root, etc.). Know in advance exactly how to do cube and other roots on your calculator! Exercises at the end of this chapter will require you to practice these skills.

Example 7.9. Solve for x from the following expression.

$$\frac{0.351}{x^3} = 9.44 \times 10^{-6}$$

Solution. Rearranging the expression to isolate x^3 on one side of the equation in the numerator:

$$\frac{0.351}{9.44 \times 10^{-6}} = x^3$$

Evaluating the fraction, we find that

$$x^3 = 3.72 \times 10^4$$

Taking the cube root, we find that

$$x = 33.4$$

You should check this with your own calculator to see if you are doing cube roots correctly. If you do not get that answer, consult your calculator manual or your instructor.

You will occasionally find cube roots in chemistry problems, but rarely roots higher than that. However, hopefully this introduction to roots will allow you to apply this knowledge should you come across them.

Exponentials and Logarithms

In considering the function "x^2," the "x" part is called the <u>base</u> and the "2" part is called the <u>exponent</u>. Such functions, generally called <u>power</u> <u>functions</u>, have the variable as the base and a number as the exponent.

Suppose you have it the other way around. Suppose a number is the base and an expression of some variable(s) is the exponent. The expression "2^x" would be an example. This expression is called an <u>exponential</u> <u>function</u>, or simply an <u>exponential</u>. (Notice the similarities in the names "exponent" and "exponential.")

The two common bases in chemistry are 10 and a number symbolized by $\underline{e}$. The number e is like π; it has a particular value of an infinite number of digits. The numerical representation of e is 2.718 281 828 46....., so you see why it is just easier to use the letter e to represent it.

We have already met some applications of exponentials. First, when we discussed scientific notation earlier, we were using the so-called "base 10" system of exponentials:

$$10^0 = 1$$
$$10^1 = 10$$
$$10^2 = 100$$
$$10^3 = 1000$$
etc.

We also found that we could take square roots, cube roots, and other roots by expressing them as fractional exponentials of numbers. As such we can calculate the square, cube, and fourth root using the **10^x** key on our calculator:

$$10^{1/2} = 3.162\ 227 \ ....$$
$$10^{1/3} = 2.154\ 434 \ ....$$
$$10^{1/4} = 1.778\ 279 \ ....$$

and so forth. In most cases, you enter the exponent into your calculator first, then press the **10^x** key. You should verify the above three roots to see if that is the case for your calculator.

The "base e" exponential , also called the <u>natural</u> <u>exponential</u>, is used the same way. There is usually a key that looks like **e^x** on most calculators. After entering the correct number for the exponent, pressing the **e^x** key will raise 2.718 281 828 46.... to that power.[†]

Example 7.10. Use your calculator to evaluate the expression

$$10^{(6.3)/(2.11)}$$

[†] The numerical value for e may seem strange, but there are definite mathematical reasons for it. Most people don't realize how many times e appears in nature (and hence its name, "natural exponential"). The curvature of the nautilus shell, the decrease in the pressure of the atmosphere with altitude, the variation of molecular speeds of gas molecules with temperature – all are related to e. Neat, huh?

> **Solution.** Remembering the proper order of operations, you should get 6.579 332 .... on your calculator. Did you?

Just as powers and roots were mathematical inverses, exponentials have inverse operations as well. They are called <u>logarithms</u>. Since we are focusing on 10 and e as bases for our exponentials, we will concentrate on the logarithms that are the specific opposite of these exponentials. They are abbreviated log and ln. (The second logarithm is understandably called the <u>natural logarithm</u>.)

It is crucial to not confuse the two logarithms. You will not get the same numerical answer when you perform a base-10 logarithm as when you perform a natural logarithm. Keep in mind that when we use the word "logarithm," we mean the **base-10** logarithm. If we use the e-related logarithm, we always state it explicitly as "**natural** logarithm."

How are logarithms related to exponentials? The logarithm of a number is that power to which the base is raised to generate that number. Pictorially,

$$\left(\text{log implies base 10}\right) \quad \log 100 = 2$$
$$10^2 = 100$$

We speak of it as "the logarithm of 100 is 2, because 10 raised to the power of 2 equals 100." Needless to say, the log 10 = 1, because $10^1 = 10$.

All positive, non-zero numbers have logarithms. (Zero and negative numbers do not.) As with roots, the logarithms of most numbers are not integers. For example, the natural logarithm of 10 is

$$\ln (10) = 2.302\ 585 \ldots$$

Remember what this means: $e^{2.302\ 585\ \cdots} = 10$. Can you show that this is true on your calculator?

Most of the time we meet exponentials and logarithms in chemistry, they are part of equations or formulas. For example, in chapter 6 we saw an equation

$$E = E^\circ - \frac{RT}{nF} \ln Q$$

where E is a voltage of a battery, E° is the battery's standard voltage, R and T have their usual definitions, n is the number of moles of electrons transferred in the balanced chemical reaction, F is the Faraday constant, and Q is what is called the reaction quotient, which is the typical "products-over-reactants" expression. In this equation, you have to evaluate the natural logarithm of the number given by the reaction quotient Q. That natural logarithm, which is a pure number with no units, is part of the equation.

One point about logarithms and exponentials: the number you take the logarithm of, or the exponent in the exponential, are **unitless**. You can only take the logarithm of a pure number. It makes no sense to evaluate "log (10 km)," since the logarithm of a unit does not exist. Similarly, when evaluating exponentials, the expression that is the exponent must be unitless overall. The next example illustrates this idea.

Example 7.11. Some exercises require that you evaluate the following expression:

$$e^{-E_A/RT}$$

where E_A is an energy, R is the ideal gas law constant, and T is the absolute temperature. Evaluate the exponential for $E_A = 25.09$ kJ/mol, $R = 8.314$ J/mol·K, and $T = 298$ K.

Solution. By substituting in the values, we see the expression becomes

$$e^{\frac{-25.09\,\text{kJ/mol}}{(8.314\,\text{J/mol·K})(298\,\text{K})}}$$

The units "mol" and "K" will cancel, but what about the energy units? They are different. One of them must be converted to another unit in order for these last units to cancel. Let us convert the kJ units into J:

$$25.09 \text{ kJ} = 25{,}090 \text{ J}$$

We substitute and get

$$e^{\frac{-25{,}090\,\text{J}}{(8.314\,\text{J})(298)}}$$

---

and now the "J" units in the expression for the exponent will cancel. We can now evaluate the exponential:

$$e^{\frac{-25,090}{(8.314)(298)}} = e^{-10.1268...} = 4.00 \times 10^{-5}$$

The final answer is a pure number; there are no units. Substitute the above numbers into your calculator and see if you get the same final answer.

Remember that logarithms and exponentials are inverses: one cancels the effect of the other. Using x as our variable, this means that

$$\log(10^x) = 10^{\log x} = x$$

Since this is so, suppose you have to find the value of an unknown that's inside a logarithm or part of an exponent of an exponential? You will have to take the exponential of both sides of the equation, or take the logarithm of both sides of the equation, in order to isolate your unknown. For example, suppose you have the following equation

$$\Delta G = -RT \ln K$$

and you are given that $\Delta G = 1.86$ kJ/mol at a temperature of 298 K and you need to find the value of the variable K. (Do not confuse it with the symbol for degrees Kelvin. Remember: know what your variables stand for!) R, the ideal gas law constant, is 8.314 J/mol·K. We substitute:

$$1.86 \text{ kJ/mol} = -(8.314 \text{ J/mol·K})(298 \text{ K}) \ln K$$

The Kelvin and mole units cancel, but the energy units are again inconsistent. Again, we will convert kJ to J and have $\Delta G = 1860$ J/mol:

$$1860 \text{ J} = -(8.314 \text{ J})(298) \ln K$$

Now we can cancel the energy units from both sides of the equation. Notice that, at this point, there are no units left.

Remember the tactic for solving for an unknown variable: it must be by itself on one side of an equation and in the numerator. If we are looking for K, we bring all of the numbers to one side of the equation to get

$$-0.751 = \ln K$$

The natural logarithm is an **operation**; we cannot simply just divide by "ln" and isolate K all by itself. This equation is asking "The natural logarithm of what number K is equal to –0.751?" In order to figure this out, we must perform the **inverse operation** in order to get rid of the logarithm. Well, the inverse operation is the exponential, and in this case it is the natural exponential e. Therefore, we raise e to both sides of the equation:

$$e^{-0.751} = e^{\ln K}$$

Since exponentials and logarithms are inverses, they cancel each other, and the right side of the equation simply equals K. Therefore,

$$K = e^{-0.751}$$

We can evaluate this using the **e^x** key on a calculator. We enter **–0.751**, press the **e^x** key, and get our final answer:

$$K = 0.472$$

Notice how we isolated K all by itself by first isolating $\ln K$ and then performing the inverse operation on both sides of the equation. Why didn't we use the **10^x** key? Because the "ln" means that we are dealing with the **natural** logarithm, not the base-10 logarithm. If you use the wrong exponential to inverse a logarithm, you will get the wrong numerical answer.

What are the rules regarding significant figures and logarithms? The most common time to consider significant figures is when you are taking a log (i.e. a base-10 logarithm) of a small number in calculating pH of aqueous solutions. The pH is defined as

$$pH = -\log [H^+]$$

where $[H^+]$ is the hydrogen ion concentration in units of molarity (which is the understood unit; it is not included inside the logarithm).

When you consider the relationship between an exponential number and its resulting logarithm, the numbers to the **left** of the decimal point of the resulting logarithm are related to the exponent of the exponential, whereas the numbers to the **right** of the decimal point of the resulting logarithm are related to the mantissa of the exponent (see Chapter 1 for the definitions of mantissa and exponent). Significant figures for logarithms are therefore limited by the significant figures of the **mantissa**. **A proper logarithm has as many decimal places as the mantissa has significant figures.** There is no restriction on the number of significant figures before the decimal point.

For example, suppose we consider the significant figures after evaluating

$$-\log (3.66 \times 10^{-6})$$

Remember that the "$\times 10^{-6}$" only places the 3, 6, and 6 in the proper decimal positions. The logarithm, negated, is

$$5.436\,518\,914\,....$$

(Is this what you get when you take the logarithm of the original number and then negate it? If not, you may be working your calculator improperly.) The number to the left of the decimal point, 5, is indicative of the exponent on the exponential and does not need to be considered. The original number had a mantissa with three significant figures. Therefore, we limit our decimal figures to three places. In this case, we round up to get for our final answer

$$-\log (3.66 \times 10^{-6}) = 5.437$$

The relationship between significant figures and decimal places is also valid when going in the opposite direction, too. For example, 10 raised to the 5.224 power is

$$10^{5.224} = 1.67 \times 10^5$$

In evaluating this, we limit our mantissa to three significant figures, because our power has three decimal places. Because of the properties of logarithms, if we took the logarithm of 1.67×10^5, we would not get 5.224 back, due to truncation errors. However, we hardly ever have to take multiple logarithms and exponentials in the same mathematical problem, so such errors do not usually affect the final answer much.

> **http://owl.thomsonlearning.com: Ch 0-4a Math: Logarithm Algebra**

> **http://owl.thomsonlearning.com: Ch 0-4b Math: Logarithm Calculations**

Student Exercises

Be careful when using your calculator! If you are not getting the correct answers, it may not be that you don't understand the math. Instead, you may not be operating your calculator correctly. If you have questions, consult your manual or see your instructor.

7.1. Write the following expressions as fractions having all positive exponents. If a variable appears more then once in an expression, be sure to combine the exponents properly.

(a) $x^5 \div y^{-3}$

(b) $\dfrac{v^3 x^{-2}}{z^4 v^{-1}}$

(c) $1 - \dfrac{y^3}{\left(\frac{1}{y^{-3}} \right)}$

(d) $\dfrac{2^8}{2^6}$

(e) $\dfrac{10^2 \cdot 3^4}{10^5 \cdot 3^{-3}}$

(f) $10^5 \cdot 10^4 \cdot 10^{-8}$.

7.2. In your home, electricity use is measured in terms of a unit called "kilowatt-hours." A similar unit composed of more basic units is the "watt-second." From the definition of the watt, write this derived unit in terms of fundamental units written with positive exponents. What is this combination of units equal to?

7.3. The units on Planck's constant, a fundamental universal constant, are J·sec. Show that this is equal to the units for angular momentum, which defined as (mass)(velocity)(distance). HINT: What are the units used to describe mass, velocity, and distance?

7.4. Determine solutions to the following quadratic equations.
(a) $x^2 - 6x + 9 = 0$.   (b) $x^2 - 16 = 0$.

(c) $a^2 - 5a + 6 = 0$.

7.5. Determine solutions to the following quadratic equations.
(a) $3x^2 + 24x - 22 = 0$.   (b) $0.8x^2 - 4.2x - 2.7 = 0$.

(c) $-6x + 6x^2 = 12$

(d) $\dfrac{4x}{0.5-x} = 10$

7.6.   Find solutions to each of the following expressions by neglecting the appropriate terms and compare your answer to solutions found using the quadratic formula. Were you justified in neglecting the terms you did?

(a) $\dfrac{2x^2}{0.0400-x} = 3.74 \times 10^{-2}$

(b) $\dfrac{x^2}{0.200-x} = 8.3 \times 10^{-5}$

7.7.   Write down the following expressions using fractional exponents and evaluate them using a calculator.

(a) The cube root of 10.

(b) The fourth root of 25.

(c) The cube root of 1000.

(d) The reciprocal of the fifth root of 250.

(e) The reciprocal of the square root of 2.

(f) The negative of the sixth root of 1.01.

7.8. Evaluate the following expressions using your calculator.

(a) $\sqrt[3]{6^{15}}$

(b) $\left(\sqrt{15}\right)^2$

(c) $\dfrac{\sqrt{16}}{\sqrt{4}}$

(d) $\dfrac{\sqrt[3]{10}}{\sqrt[4]{10}}$

7.9. Solve the following equations for the unknown variable.

(a) $\dfrac{56.0}{x^3} = 1.44 \times 10^{-2}$

(b) $\dfrac{4977}{x^3} = \dfrac{22\,844}{x^5}$

(c) $x = e^{-569/444}$

(d) $2.964\,23 = 10^x$

(e) $2.1116 \cdot \ln\left(\dfrac{x^2}{0.654}\right) = -0.4482$

(f) $e^{\ln 8.883} = x$

(g) $\log\left(\dfrac{10^3 \cdot 10^{-5}}{10^4 \cdot 10^2}\right)$

(h) $1.077 = 1.100 - \dfrac{(8.314)(295)}{(2.00)(96,500)}\ln Q$

7.10. Most people don't memorize the value for e, since it is easy to get from your calculator. Can you get your calculator to display the value for e? There are several ways, depending on the model of calculator you have.

7.11. What is the pH of a solution that has $[H^+] = 4.208 \times 10^{-3}$? Use the equation $pH = -\log[H^+]$, and express your answer to the proper number of significant figures.

7.12. What is the $[H^+]$ of a solution that has a pH of 10.882? Express your answer to the correct number of significant figures.

7.13. Throughout the course of this book, we have discussed several mathematical operations that are "opposite" or "the inverse" of each other. Name four pairs of operations that can be considered "opposites."

Answers to Student Exercises

7.1. (a) $x^5 y^3$ (b) $\dfrac{v^4}{z^4 x^2}$ (c) 0 (d) 2^2, or 4 (e) $\dfrac{3^7}{10^3}$, which equals 2.187 (f) 10^1, or 10.

7.2. A watt-second equals $\dfrac{kg \cdot m^2}{sec^2}$, which is equal to a joule. Watt-seconds are therefore a measure of energy.

7.3. Both sets of units can be shown to equal $\dfrac{kg \cdot m^2}{sec}$.

7.4. (a) $x = 3$ (b) $x = 4$ and $x = -4$ (c) $a = 2$ and $a = 3$.

7.5. (a) $x = 0.830\ 458 \ldots$ and $x = -8.830\ 458 \ldots$ (b) $x = 5.829 \ldots$ and $x = -0.579 \ldots$ (c) $x = 2$ and $x = -1$. (d) $x = 0.427 \ldots$ and $x = -2.927 \ldots$

7.6. (a) Neglecting the x in the denominator of the original expression, we find $x = \pm\, 0.027\ 349 \ldots$, while using the quadratic formula, we find $x = 0.019\ 553 \ldots$ and $x = -0.038\ 253 \ldots$ Neglecting the x does not give a good approximate answer. (b) Neglecting the x: $x = \pm\, 4.074\ 309 \ldots \times 10^{-3}$, while using the quadratic formula, $x = 4.033\ 021 \ldots \times 10^{-3}$ and $x = -4.116\ 021 \ldots \times 10^{-3}$. This approximation was justified.

7.7. (a) $10^{\frac{1}{3}} = 2.154\ 434 \ldots$ (b) $25^{\frac{1}{4}} = 2.236\ 067 \ldots$ (c) $1000^{\frac{1}{3}} = 10$ (d) $\dfrac{1}{250^{\frac{1}{5}}} = 250^{-\frac{1}{5}} = 0.331\ 445 \ldots$

(e) $\dfrac{1}{2^{\frac{1}{2}}} = 2^{-\frac{1}{2}} = 0.707\ 106 \ldots$ (f) $-\left(1.01^{\frac{1}{6}}\right) = -1.001\ 659\ 76$.

7.8. (a) $2.449\ 489\ 742 \ldots$ (b) 15 (c) 2 (d) $1.211\ 527 \ldots$

7.9. (a) 15.7 (to three sig figs) (b) ± 2.142 (four sig figs) (c) 0.278 (d) 0.471 91 (e) 0.727 (273 973 ...) (f) 8.883 (g) -8 (h) $Q = 6.11$

7.10. To get the value for e, enter **1** into your calculator and hit the e^x key. If your calculator does not have an e^x key, you may have to hit the inverse natural logarithm, probably **INV**, then **ln**. Which way works for your calculator?

7.11. pH = 2.3759.

7.12. $[H^+] = 1.31 \times 10^{-11}$, where the unit is assumed to be molarity.

7.13. The following pairs of operations can be considered opposites or inverses of each other:

addition & subtraction

multiplication & division

powers and roots

logarithms and exponentials

Chapter 8. Making Graphs

Introduction

Graphs are commonly used in chemistry and other sciences to visually represent related sets of information. In a glance, they can illustrate information in a way that tables of data can't, and often they are the best way to illustrate trends and relationships between different observable parameters of a chemical system. Many times – especially in a laboratory course – you will have to take data that you yourself measured and plot them on a graph to see what trend or relationship exists. It is important that the graph be represented properly for the best possible understanding of the information. A bad graph, with data poorly plotted, is a waste of paper more than it is an effective tool for communicating information.

In this chapter, we will review some of the basic concepts for constructing good graphs. The way to set up a graph depends on what you want it to show, and there are a few choices you will have to make in order to present your data effectively or to determine additional information from your graph. Because of the wide range of possibilities for setting up a graph, we will focus on one major example and develop it, instead of presenting multiple examples throughout the chapter.. Hopefully, you will be able to apply the ideas for constructing proper graphs to the student exercises by the end of the chapter.

Straight or Curved Lines?

When you plot data on a graph, you are using data that are related to each other somehow, and the graph is a visual representation of that relationship. For example, if you have a fixed amount of gas at a certain temperature, the volume of that gas is related to its pressure. (This relationship is called Boyle's law.) If you made measurements of the pressure of gas at different volumes, you could plot the different values of pressure versus the volume and produce a graph.

In almost all cases of related data, the change in one variable with respect to the other is **smooth**. That is, there are few abrupt and sudden changes, unless something abrupt is done to your system (like dropping a hot piece of metal into cool water and measuring the resulting temperature of the water). What this means is that most of your graphs will look like the following two:

Two related variables can also be plotted to give a curved line. One example is the relationship between pressure and volume of a fixed amount of gas at a given temperature. The mathematical equation that relates pressure P and volume V is

$$P = K \cdot \frac{1}{V}$$

where K is a proportionality constant. This equation suggests that a plot of P versus $\frac{1}{V}$ would give a straight line, and it does. But it is more common to plot P versus V directly, and when you do, you get a graph that looks like

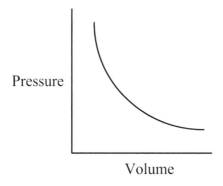

In this case, the graph is a smooth curve, not a straight line. (Notice that we have labeled our axes for the first time. We will be discussing this shortly.)

In short, it's a good idea to know ahead of time whether your graph is expected to be a straight line or a curved line. No graph, however, is ever going to be a **perfect** straight or curved line. That's because when you are working with experimental data, there will always be experimental error. This experimental error usually adds a little bit of imperfection, or <u>scatter</u>, to your graph. But the trend in the data plotted in a graph should still be obvious: either a straight line or a curved line.

Being able to recognize that an equation should yield a straight line is an important skill. The following example gives some idea of how to spot equations that can be graphed to give a straight line.

Example 8.1. Each of the following equations will yield a straight line, if plotted correctly. For each given set of variables, indicate what to plot as y, what to plot as x, and identify the correct expression for m and b. Assume that all other variables are constants. (a) $\ln A - \ln K = -kt$, where the variables are

A and t. (b) $\dfrac{1}{A} = \dfrac{1}{K} + kt$, where the variables are A and t. (c) $pH = pK + \log(\text{ratio})$, where pH and "ratio" are the variables.

Solutions. (a) If we rewrite the equation to

$$\ln A = -kt + \ln K$$

then we would plot $\ln A$ as y and t as x. The slope would be $-k$, while the y-intercept would be $\ln K$. (b) For this equation, $\dfrac{1}{A}$ would represent y, t would be our x, the slope would be k, and the y-intercept would be $\dfrac{1}{K}$. (c) The pH would be the y variable, and the "log(ratio)" would act as the x variable. There is no visible constant multiplying the "log(ratio)" term, but there is always an understood 1 as a multiplier. Therefore, the slope of this equation is equal to 1. The y-intercept will be equal to pK.

Axes

When graphing the behavior of one variable versus another, you commonly start out with a set of data, usually in the form of a table. Usually, one of the variables listed in the table has values that were determined by the experimenter, and the other variable is the property that was measured. The variable that is determined is called the independent variable, while the variable that is measured is the dependent variable. Graphs are plots of how a dependent variable changes with respect to the independent variable.

Most graphs are set up by using two perpendicular scales, one for each variable, called axes (singular axis). Some graphs use a circular grid to plot data, but all of the cases we will be considering will use two perpendicular axes. Typically, the horizontal axis is considered the x axis, and the independent variable is measured against this axis. The vertical axis is referred to as the y axis, and the dependent variable is measured against this axis:

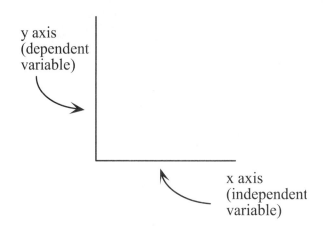

If you use graph paper you bought in a store, you will probably notice that each page isn't completely covered with a grid of lines (which helps in plotting your data). Some of the lines of the grid are usually darker than others. This helps you keep track of your position. We will start presenting graphs on a grid and will do so for the remainder of this chapter.

When you set up a graph, you should first draw the axes yourself with a pen or pencil. You can either draw them on the edge of the grid of lines, or you can indent your graph by drawing axes independent of the grid's edge. The overall look of your graph should not be affected by exactly where you draw your own axes – as long as you give yourself an appropriate amount of space for all of your data. (WARNING: Your instructor may require that you draw your axes in a particular way. You should check with your instructor regarding the placement of axes on a sheet of graph paper.) For example, you might have a set of axes that look like this:

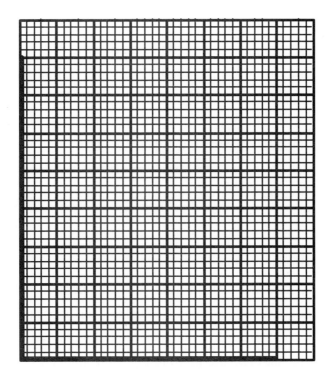

Or, you may choose to "indent" your axes so they aren't on the edge of the grid:

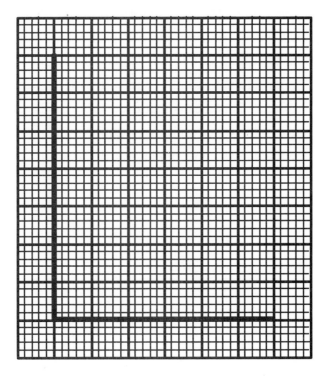

Again, you should check with your instructor to find out whether one way of drawing axes is preferred. Many graph papers have grids showing dark and light lines, like the examples above. Each dark line in these examples is five marks away from the next dark line, and four lighter lines separate them. If the axes you draw are indented, they are usually drawn over the darker grid lines on the graph paper. (Graph paper comes with grids marked out in English units, metric units, and even logarithmic units, but they all usually have dark and light grid lines to help keep track of your position along an axis.) Axes sometimes represent a zero value for one of the variables. If this is the case for your particular graph and some of your data are negative numbers, you may have to indent your axes in order to accommodate all of the data in your plot.

Numbering and Labeling Axes

For beginning students, it seems that the most difficult part of setting up a graph is how to number and label the axes, in order to give each axis a scale. This is because the scales for the axes of every graph depend on the data, and when different data are being plotted, the scale for that graph will be different. What this means is that there are no truly absolute rules for determining the scales for the axes. However, there are some general guidelines that will make any graph a more acceptable visual presentation of your data.

As mentioned, the grids on graph paper can be marked out in metric units or English units. In both cases, the intervals between the consecutive parallel grid lines are the same. These are linear scales, and each interval is meant to represent the same change in the variable's value. There is also a logarithmic scale, where each major interval (usually marked by the darker grid lines or slightly larger tick marks) represents a power of ten. This means that each successive major interval is ten times the previous one. This allows you to plot small and large numbers on the same axis, but the scale is completely different. The following shows an example of a linear scale versus a logarithmic scale for one axis:

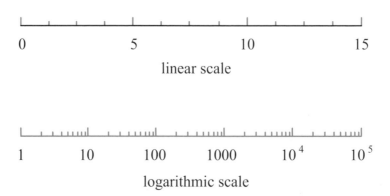

If both axes are logarithmic, the graph is referred to as a "log-log graph." In chemistry we almost exclusively use linear plots.

Good graphs should use <u>most of the area bound by the axes</u>. Furthermore, the area bound by the axes should use most of the area of the page of graph paper. These two ideas allow you to make the most efficient use of space for the graph. The axes drawn on the previous page use almost all of the area of the grid on the graph paper. As a poor example, consider the following:

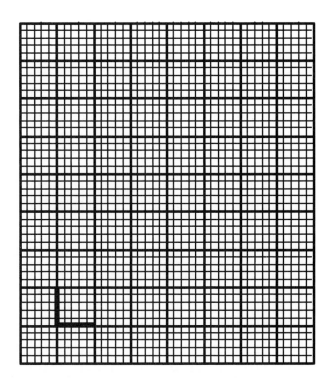

In this case, axes are drawn but only cover a fraction of the total area of the available graph paper grid. If you were to try to plot a collection of points in such a small area, any trend in your data would be hard to see.

Drawing the axes for a graph and determining what values each grid line should have cannot be done without consideration of the data. Most axes start and end on a dark grid line, if the graph paper has them. These darker lines are the lines of the grid that are usually marked to indicate specific values of the variable being plotted on that axis. The extreme values represented by the first and last interval <u>should include the entire range of values</u> for the dependent and independent variables. Each major interval (i.e. the distance between labeled lines) <u>should have the same change in value</u> (for linear graphs). This means that you have to consider your data, the size of your graph, and the number of major intervals each axis will have. Usually each interval has a simple change in value, like 1 or 3 or 5 or 10. Good graphs don't have intervals having odd values, like 5.77 or 0.054 22 or 10.6209.

As an example, consider the following data collected for a gas at constant volume:

Temperature, K	Pressure, atm
100	0.200
200	0.400
250	0.500
300	0.600
350	0.700
400	0.800

Suppose we want to plot the independent variable temperature versus the dependent variable pressure. Temperature is on the x axis and pressure is on the y axis. Temperature ranges from 100 K to 400 K, and the pressure ranges from 0.200 atm to 0.800 atm. Suppose, too, that we are working with a piece of graph paper like so:

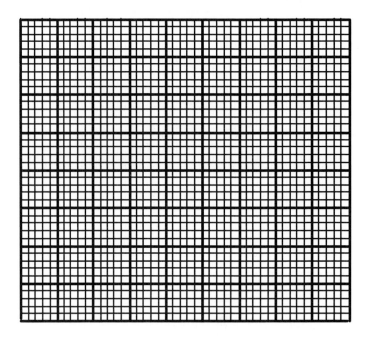

We have a total of nine major intervals for our x axis, which will be for temperature, and a total of 8 major intervals for our y axis, which will be for pressure. There are many possible ways of setting up this graph; what follows is not the only way. Consider the y axis first. If each major interval were equal to 0.1 atm, then we can use the edge of the grid and label each dark line from 0 to 0.8 atm in 0.1-atm increments. We will be able to include all of our pressure data on that scale. Our y axis for our graph will look like this:

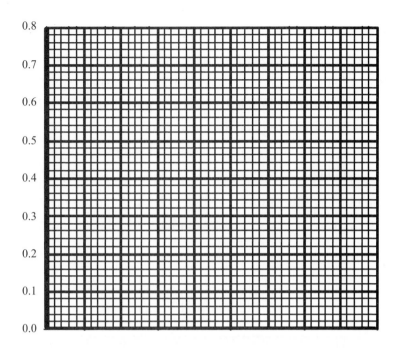

For the x axis, if we have each major interval represent 50 K, we can go from 0 K to 450 K. If we include our x axis, the graph would now look like this:

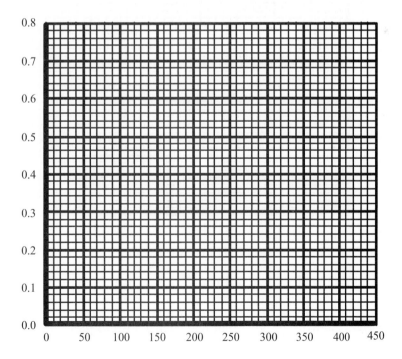

While it may seem that we are now ready to plot the data points and finish our graph, there is something missing. Since a graph is supposed to be a concise pictorial representation of data, the graph must include what the numbers stand for. This is crucial, because graphs are useful for illustrating principles to other people, and these other people won't necessarily know what is being plotted with respect to what! **Each axis of the graph must be labeled with the appropriate variable <u>name</u> and <u>unit</u>.** Both variable name and unit are necessary, especially since many variables can be expressed in different units. In this case, the x axis is the temperature in units of degrees Kelvin, and the y axis represents pressure in units of atmospheres. Many times, a graph is also given a title so that people can differentiate between graphs. The title is often written outside the grid on the page if there is room, or directly on the grid out of the way of any of the data points. The labels and title must be included in a proper graph, like so:

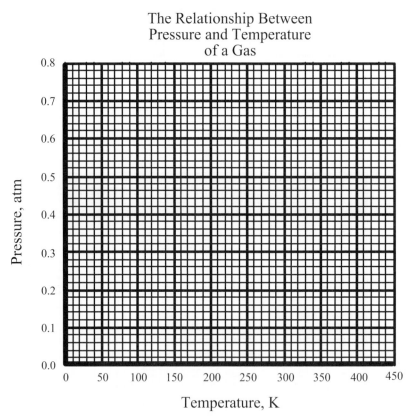

We are now ready to take our data and construct the plot.

Example 8.2. Shown below is a different grid that has been set up to plot the temperature/pressure data from the table above. Although the grid is somewhat smaller than the example we worked out above, there are some deficiencies in how the axes were set up. What do you think could be done better?

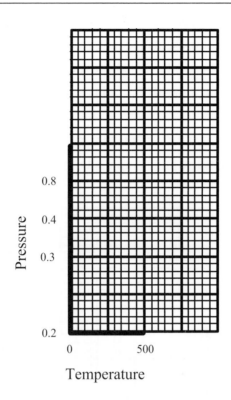

Suggestions for Improvement: The *x* axis is drawn too short, and the temperature values won't be spread out enough for any relationship to be noticeable. There are no units on either axis. The *y* axis doesn't extend as far as it could, and the numerical labels for the major intervals aren't equally spaced. The *y* axis could be extended so that each interval is 0.1 and all of the pressure data could be plotted. The graph does not have a title. There are many possible ways to set up a better representation to graph the given data.

Plotting the Points

Now that the axes for the graph have been set up, the individual data sets can be plotted. Let us recall the data we are plotting:

Temperature, K	Pressure, atm
100	0.200
200	0.400
250	0.500
300	0.600
350	0.700
400	0.800

You need to recognize which numbers make up a "data set." Our two measurements are pressure versus temperature, so each line in the table represents a set of two numbers that go together. If we wanted to express each pair of numbers as (T, P), we have (100 K, 0.200 atm), (200 K, 0.400 atm), etc. Each of these sets of numbers is represented by a particular point in our two-dimensional graph. What we do is draw a dot on the grid for each data set.

Consider the first data point, (100 K, 0.200 atm). Locate 100 K on the temperature axis and 0.200 atm on the pressure axis. (The title of our graph is going to be omitted for sake of clarity.)

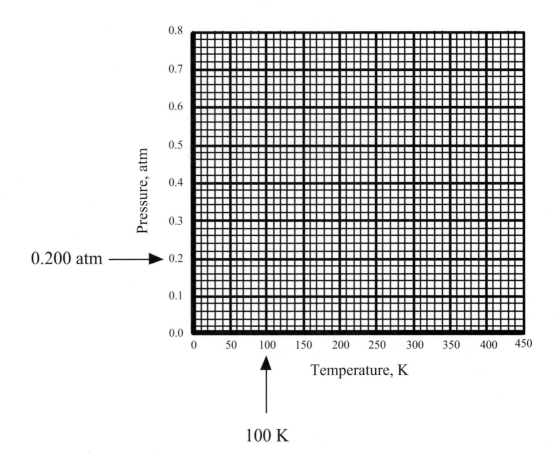

At each value, draw lightly on the graph (or imagine in your mind) a line, starting from each value at the respective axis and into the grid **parallel to the other axis**. The two lines should intersect:

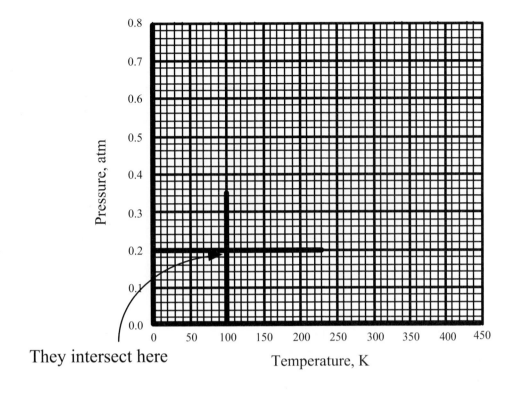

They intersect here

Temperature, K

That point of intersection is the point on your graph that represents the data set (100 K, 0.200 atm). Draw a dot at that point:

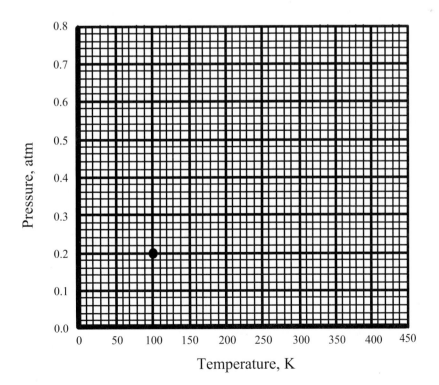

Temperature, K

The other five data sets can be plotted in a similar way. When finished, our graph looks as follows:

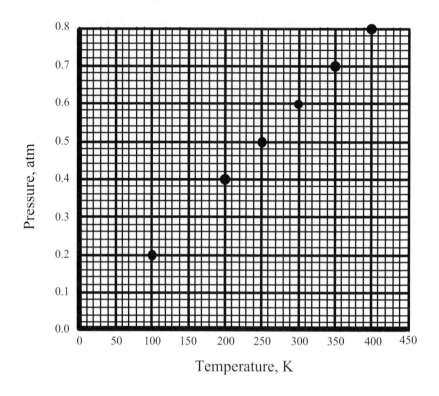

Although the dots used here seem large, they are that size so that they are obvious. You should use as small a dot as you can. Some instructors ask that you circle your tiny dots, so they are more obvious to another person who might be reading the graph.

Drawing the Trend

Remember it was mentioned earlier that almost all graphs illustrate relationships that are smooth. That means that we can establish the overall trend of our measurements with only a limited number of data points, and we can do so with a high degree of confidence. For example, the points that were plotted in the above graph seem to be in a straight line. Even though we do not have measurements for every single point along that line (and we never will, since a line is composed of an infinite number of points), we can confidently predict that if we were to ever measure every single point, all of the points plotted on the graph would make a straight line. In graphing, however, we discover the trend by making only a few measurements, indicated by the dots, and then **smoothly** drawing a straight line through the dots to indicate the overall trend:

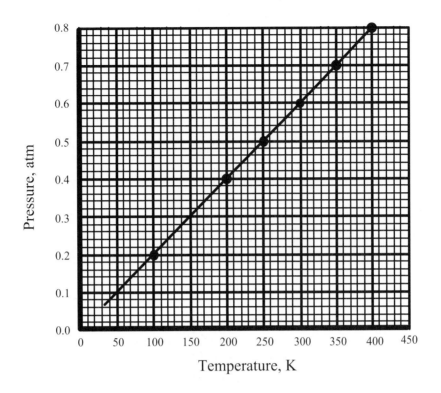

This graph shows that the points are in a straight line. Using this graph, we can understand two things. First, we see that there is a linear relationship between T and P. Second, we can use the graph now to predict the pressure of our gas sample at a temperature **that is not measured experimentally**. For example, what pressure would the gas have at 150 K? The plot shows that "150 K" will intersect with "0.3" atm, and that this is the predicted pressure of the gas.

In reality, the points on the plot won't be this perfect, owing to experimental error. In any real measurement, there will be some uncertainty and some error, and the data set won't be so perfect when plotted. This does not mean that your data are wrong, or that you will not be able to construct a good graph of your data! Remember, you should have an idea of how your data should plot in the beginning, and if you expect a straight line, then the non-ideality of your points may be more a reflection of experimental non-ideality instead of a non-linear (or non-curved) relationship. When a situation like this occurs, how do we draw the graph? **We do not simply connect the dots!** (We left that game behind in kindergarten.) Instead, we make a "best guess" fit. We draw a line that is a best approximate fit to all of the points, but not a perfect fit to any of the points. For example, if you made temperature and pressure measurements for a real gaseous system and plotted your points, they might have some scatter to them, and a more realistic graph might look as follows:

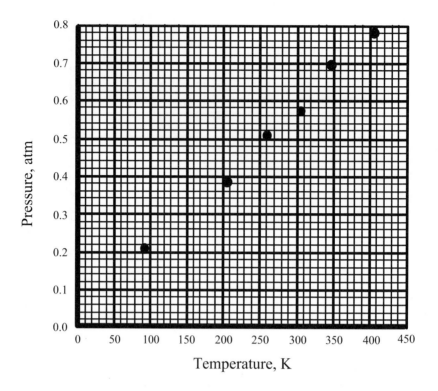

The points are not in a perfect straight line. However, we can "eyeball" a line passing very close to these points, and this line should be a good general representation of the behavior of our system:

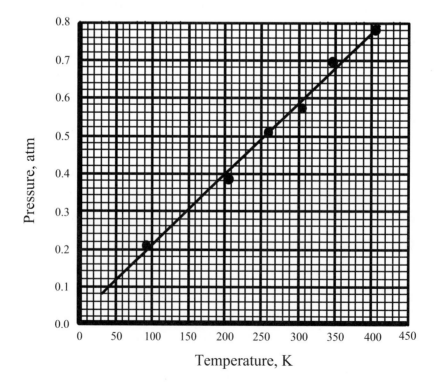

Notice that not all of the points lie directly on the line; some are above and some are below. Usually, about half of the points will be above the line, and about half will be below the line. (This is not always the case, because a "best fit" line also takes into consideration how far the data point is from the line.) This is one example of what would be called a "best fit" line that describes all of these data points.

A true "best fit" line can actually be calculated mathematically, and some calculators can be programmed to determine the mathematically-optimal, true best fit line. However, we won't go into that in this chapter. In most cases in the initial study of chemistry, an "eyeball" best fit line is acceptable. Your instructor should inform you if you are required to use a calculator or computer to calculate a mathematical best fit line for a graph.

Of course, some data will give a curved line, not a straight line. The issues of smooth changes, non-ideal behavior of data, and "best fit" lines also apply. For example, you might have a graph of volume versus pressure that looks like this:

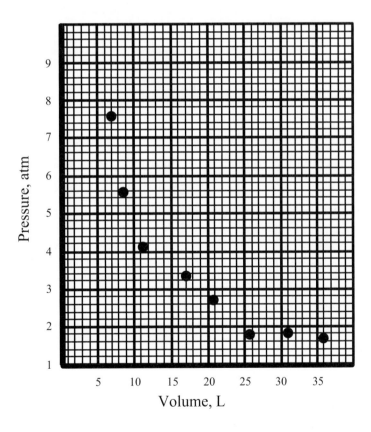

We expect that there is a curved relationship between volume and pressure, and so we can "eyeball" a best fit curved line to represent the smooth change in pressure as the volume changes:

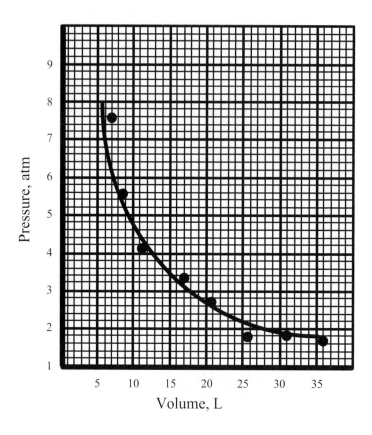

Drawing best fit lines for plotted data in a graph takes practice. In time, you will develop the self-confidence necessary to convince yourself that your ability to guess the best fit line for a set of imperfect experimental data is reasonable. One suggestion is to compare your best fit graphs with the graphs of others using similar data to see how they graph the same information.

---

http://owl.thomsonlearning.com: Ch 0-3b Math: Functions and Graphs/Basics

---

http://owl.thomsonlearning.com: Ch 0-4c Math: Functions and Graphs/Logs and Exponents

---

Extrapolation

One aspect of graphing involves projecting the line into regions where data do not exist and then predicting values of measurables instead of measuring them directly. This process is called extrapolation. Extrapolation is almost always done with data that are expected to yield a straight line, because it is very easy to extend a straight line as far as you want.

One of the more common uses of extrapolation on graphs in chemistry is to determine the actual value of a y-intercept from data that should make a straight line. For example, in the equation

$$\ln k = \ln A - \frac{E_A}{RT}$$

the "ln A" term is the y-intercept. Because our x axis variable is defined as $\frac{1}{T}$ (see above), we will not get to $\frac{1}{T} = 0$ unless $T = \infty$ (which is physically impossible). Thus, we determine the y-intercept by extrapolation.

The key to extrapolating to a y-intercept is that **the x axis of the graph must go to zero**. Why? Because the true y-intercept of a straight line occurs at $x = 0$. If x, or whatever expression that is plotted on the x axis, does not have a zero point on the scale, you will not be able to determine the y-intercept by extrapolation.

Consider the following set of data:

k	$\ln k$	T, K	$1/T$, K^{-1}
1.18×10^{-5}	-11.347	293	3.41×10^{-3}
2.35×10^{-5}	-10.659	298	3.36×10^{-3}
4.67×10^{-5}	-9.972	303	3.30×10^{-3}
9.10×10^{-5}	-9.305	308	3.25×10^{-3}
1.81×10^{-4}	-8.617	313	3.19×10^{-3}

If we were going to plot $\ln k$ on the y axis versus $\frac{1}{T}$ on the x axis, we would get a graph that looks like this:

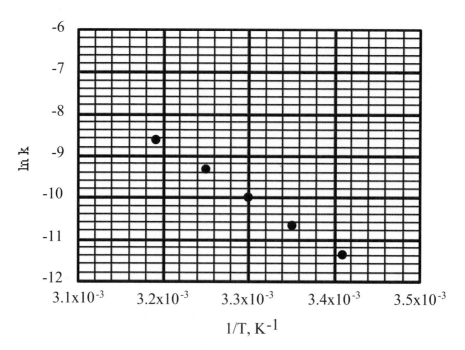

Could we simply draw a best fit line and extrapolate to the edge of the grid to determine the *y*-intercept? **NO!** The *x* axis does not go all the way to zero, which is where the true value of the *y*-intercept is. If we replot the graph, and make the minimum value of the scale equal to zero, we can extrapolate on that graph to the proper *y*-intercept. Of course, if we change one scale, we may have to change the other scale so that we can determine an approximate value for the *y*-intercept. That is the case in this example. But in doing so, we can extrapolate a value for the *y*-intercept by using these data, and in doing so graphically determine an approximate value for ln A. The graph looks like this:

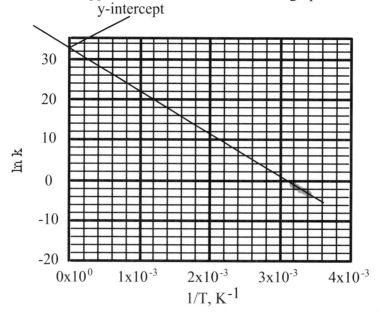

Notice how all of our data are clustered in one section of the graph, and how the plotted points take up only a small portion of the graph. This may be cause for some concern. The relationship between the *x*-axis variable and the *y*-axis variable may not be perfectly linear over the entire interval. Large extrapolations like this may be suspect because the exact *y*-intercept depends not only on the reliability of the original data, but also the appropriateness of our best fit line. If there is too much scatter in the experimental data, or if a poor best fit line is drawn, then extrapolating the straight line for such a distance could induce a major uncertainty in the value of our *y*-intercept. Consider the two lines below:

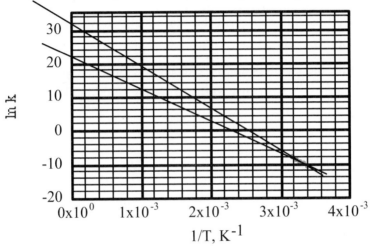

Which line provides the "better" *y*-intercept? One line gives $\ln A \approx 22$, while the other gives $\ln A \approx 34$. Working backwards, these give values of A approximately equal to 3.5×10^9 or 5.8×10^{14}, respectively. A five order of magnitude difference for a variable is not inconsequential! This difference illustrates the precautions you need to take when performing extrapolations of your data.

Student Exercises

Only a few exercises are given here because there are so many aspects to graphing data and the process is always open to a little personal interpretation. Because of this, no "answers" are given for these exercises. If you wish to check the appropriateness of your graph, ask another student or your instructor to comment on your work for you. Several pages of graph paper are included after the exercises, or you can purchase your own. Your instructor may want you to use a certain style of graph paper, so you should check with your instructor.

8.1. Construct a graph of the following data, which relates the density of water in g/cm^3 to the temperature. Draw a best fit curve for the points.

Temperature, °C	Density	Temperature, °C	Density
0	0.9998	50	0.9880
10	0.9997	60	0.9832
20	0.9982	70	0.9778
30	0.9957	80	0.9712
40	0.9922	90	0.9654
		100	0.9584

8.2. Various coefficients are used to represent the non-ideal behavior of real gases. A common one is called the second virial coefficient and has the symbol B. B varies with temperature, and at the temperature where B is zero, the gas acts almost exactly like an ideal gas. Graph B versus temperature for carbon dioxide, CO_2, draw a best fit line to illustrate the trend, and estimate the temperature where CO_2 acts like an ideal gas.

Temperature, K	B
400	−62
500	−30
600	−13
700	−1
800	7
900	12
1000	16
1100	19

8.3.   Hydrocarbons are simple molecules composed of carbon and hydrogen.   The simplest hydrocarbons are long chains of carbons with hydrogens attached to the long chain.   There is an interesting trend in the melting points of these hydrocarbons: the greater the number of carbon atoms in the chain, the higher the melting point. Use the data in the table below to extrapolate and predict the number of carbon atoms in the molecule that melts around 0°C. Can you look up this information in another reference and check your prediction?

# of C atoms in molecule	Melting point, °C
17	22.0
18	28.2
19	32.1
20	36.8

8.4. A student is making a measurement of the solubility of a salt in a NaCl (sodium chloride) solution. The following data are collected experimentally:

NaCl concentration, M	Solubility of second salt, g/L
0.05	1.7×10^{-6}
0.025	9.5×10^{-7}
0.0125	5.1×10^{-7}
0.00625	2.9×10^{-7}

Graph the solubility versus the NaCl concentration and extrapolate to determine the solubility of the salt in pure water, i.e. where the concentration of NaCl $= 0$.

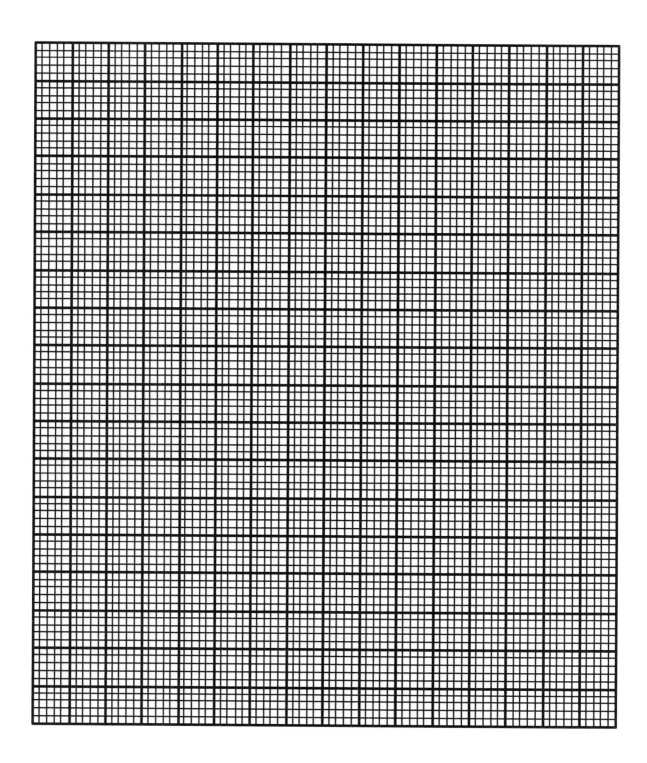

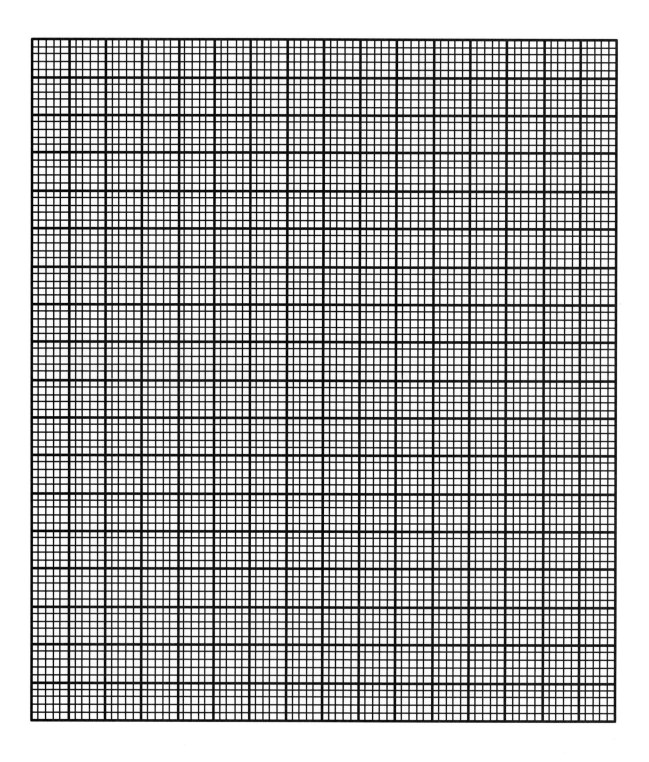

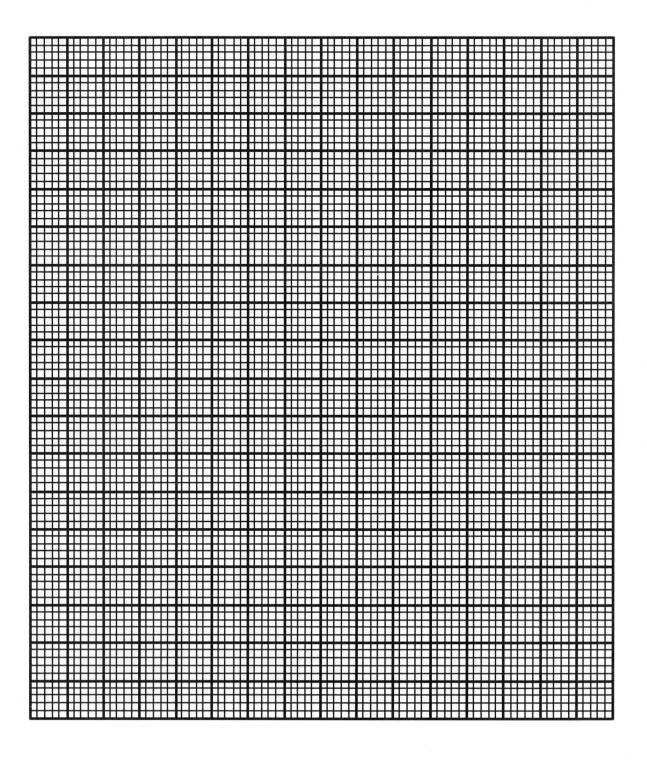

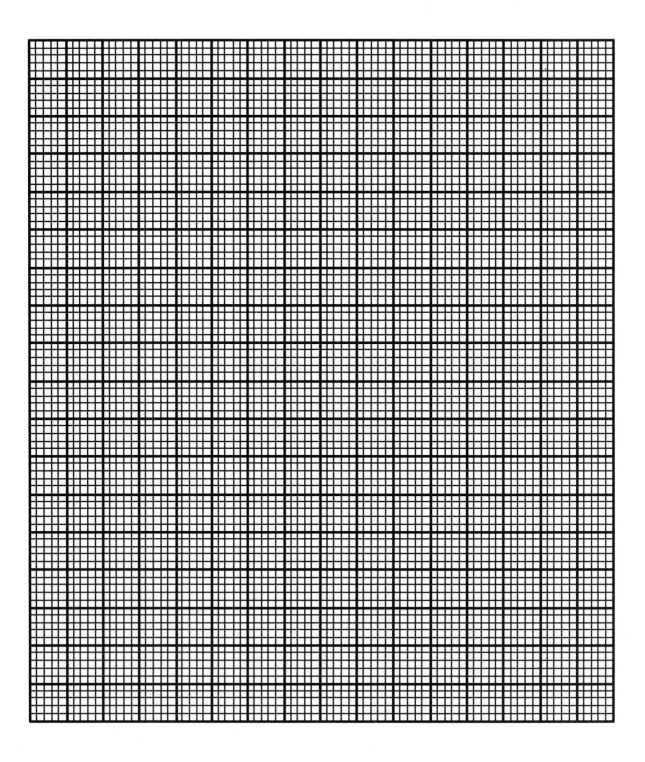

Computer Basics

Windows® 10 Edition

Michael Miller

QUE®

800 East 96th Street
Indianapolis, Indiana 46240

CONTENTS

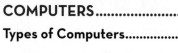

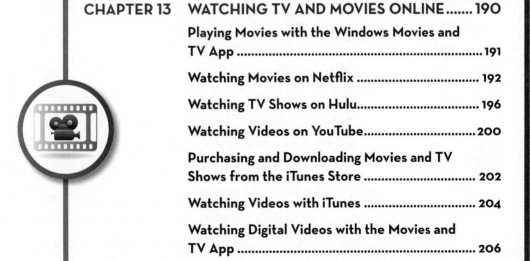

EASY COMPUTER BASICS, WINDOWS® 10 EDITION

ISBN-13: 978-0-7897-5452-3
ISBN-10: 0-7897-5452-5

Library of Congress Control Number: 2015945675

Printed in the United States of America

First Printing: September 2015

TRADEMARKS

WARNING AND DISCLAIMER

SPECIAL SALES

For information about buying this title in bulk quantities, or for special sales opportunities (which may include electronic versions; custom cover designs; and content particular to your business, training goals, marketing focus, or branding interests), please contact our corporate sales department at corpsales@pearsoned.com or (800) 382-3419.

For government sales inquiries, please contact governmentsales@pearsoned.com.

For questions about sales outside the U.S., please contact international@pearsoned.com.

Associate Publisher
Greg Wiegand

Acquisitions Editor
Michelle Newcomb

Development Editor
Charlotte Kughen

Managing Editor
Kristy Hart

Senior Project Editor
Betsy Gratner

Technical Editor
Vince Averello

Copy Editor
Cheri Clark

Indexer
Lisa Stumpf

Proofreader
Leslie Joseph

Publishing Coordinator
Cindy Teeters

Compositor
Mary Sudul

ABOUT THE AUTHOR

Michael Miller is a successful and prolific author with a reputation for practical advice, technical accuracy, and an unerring empathy for the needs of his readers.

Mr. Miller has written more than 150 best-selling books over the past 25 years. His books for Que include *Easy Facebook*, *My Facebook for Seniors*, *My Social Media for Seniors*, *My Windows 10 Computer for Seniors*, and *Computer Basics: Absolute Beginner's Guide*.

He is known for his casual, easy-to-read writing style and his practical, real-world advice—as well as his ability to explain a variety of complex topics to an everyday audience.

Learn more about Mr. Miller at his website, www.millerwriter.com. Follow him on Twitter @molehillgroup.

DEDICATION

To Sherry—life together *is* easier.

ACKNOWLEDGMENTS

Thanks to the usual suspects at Que, including but not limited to Greg Wiegand, Michelle Newcomb, Charlotte Kughen, Cheri Clark, Betsy Gratner, and technical editor Vince Averello.

WE WANT TO HEAR FROM YOU!

As the reader of this book, *you* are our most important critic and commentator. We value your opinion and want to know what we're doing right, what we could do better, what areas you'd like to see us publish in, and any other words of wisdom you're willing to pass our way.

As an associate publisher for Que Publishing, I welcome your comments. You can email or write me directly to let me know what you did or didn't like about this book—as well as what we can do to make our books better.

Please note that I cannot help you with technical problems related to the topic of this book. We do have a User Services group, however, where I will forward specific technical questions related to the book.

When you write, please be sure to include this book's title and author as well as your name, email address, and phone number. I will carefully review your comments and share them with the author and editors who worked on the book.

Email: feedback@quepublishing.com

Mail: Que Publishing
ATTN: Reader Feedback
800 East 96th Street
Indianapolis, IN 46240 USA

READER SERVICES

Visit our website and register this book at informit.com/register for convenient access to any updates, downloads, or errata that might be available for this book.

IT'S AS EASY AS 1-2-3

Each part of this book is made up of a series of short, instructional lessons, designed to help you understand basic information.

1 Each step is fully illustrated to show you how it looks onscreen.

2 Each task includes a series of quick, easy steps designed to guide you through the procedure.

3 Items that you select or click in menus, dialog boxes, tabs, and windows are shown in **bold**.

Tips, notes, and cautions give you a heads-up for any extra information you might need while working through the task.

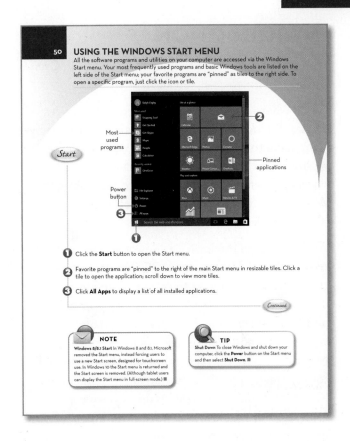

50

USING THE WINDOWS START MENU

All the software programs and utilities on your computer are accessed via the Windows Start menu. Your most frequently used programs and basic Windows tools are listed on the left side of the Start menu; your favorite programs are "pinned" as tiles to the right side. To open a specific program, just click the icon or tile.

Most used programs

Start

Power button

Pinned applications

1 Click the **Start** button to open the Start menu.

2 Favorite programs are "pinned" to the right of the main Start menu in resizable tiles. Click a tile to open the application; scroll down to view more tiles.

3 Click **All Apps** to display a list of all installed applications.

Continued

NOTE

Windows 8/8.1 Start In Windows 8 and 8.1, Microsoft removed the Start menu, instead forcing users to use a new Start screen, designed for touchscreen use. In Windows 10 the Start menu is returned and the Start screen is removed. (Although tablet users can display the Start menu in full-screen mode.) ■

TIP

Shut Down To close Windows and shut down your computer, click the **Power** button on the Start menu and then select **Shut Down**. ■

INTRODUCTION

Computers don't have to be scary or difficult. Computers can be *easy*—if you know what to do.

That's where this book comes in. *Easy Computer Basics, Windows 10 Edition* is an illustrated, step-by-step guide to setting up and using your new computer. You'll learn how computers work, how to connect all the pieces and parts, and how to start using them. All you have to do is look at the pictures and follow the instructions. Pretty easy.

After you learn the basics, I show you how to do lots of useful stuff with your new PC. You learn how to use Microsoft Windows to copy and delete files; use Google's Gmail and the Windows Mail app to send and receive email messages; use Microsoft Word to write letters and memos; use your web browser and Microsoft's new Cortana virtual assistant to search for information on the Internet; and use Facebook, Pinterest, and Twitter to keep up with what your friends are doing. We even cover some fun stuff, including listening to music, viewing digital photographs, and watching movies and TV shows online.

If you're worried about how to keep your PC up and running, we cover some basic system maintenance too. And, just to be safe, I also show you how to protect your computer when you're online. It's not hard to do.

To help you find the information you need, I've organized *Easy Computer Basics, Windows 10 Edition* into 16 chapters.

Chapter 1, "Understanding Personal Computers," discusses all the types of personal computers out there and describes the pieces and parts of a typical computer system. Read this chapter to find out all about desktops, all-in-ones, notebooks, and tablets—and the things like hard drives, keyboards, mice, and printers that make them tick.

Chapter 2, "Setting Up Your PC," shows you how to connect all the pieces and parts of a typical PC and get your new computer system up and running.

Chapter 3, "Connecting Peripherals and Other Devices," shows you how to connect other things—including your living room TV—to your new computer.

Chapter 4, "Setting Up a Wireless Home Network," helps you connect all the computers in your house to a wireless network and share a broadband Internet connection.

Chapter 5, "Using Microsoft Windows 10," introduces the backbone of your entire system—the Microsoft Windows 10 operating system—and shows you both basic operations and how to get the most out of it.

Chapter 6, "Personalizing Windows," shows you how to customize Windows 10's desktop and lock screen, how to change colors and backgrounds, and how to add new users to your system.

Chapter 7, "Working with Software Applications," walks you through everything you need to know about software applications, including how to find new apps in Microsoft's Windows Store.

Chapter 8, "Using Microsoft Word," shows you how to use both desktop and online versions of Microsoft's popular word processor to create letters and other documents.

Chapter 9, "Working with Files and Folders," shows you how to use File Explorer and Microsoft's OneDrive to manage all the computer files you create—by moving, copying, renaming, and deleting them.

Chapter 10, "Using the Internet," is all about how to get online and what to do when you're there—including how to use the new Edge web browser to surf the Web, search for information, and shop for items online. You'll also learn how to use Cortana, Microsoft's virtual assistant, to search both the Internet and your own computer.

Chapter 11, "Communicating with Email," is all about using email to communicate with friends, family, and co-workers. The focus is on Windows 10's Mail app, as well as Google's Gmail service.

Chapter 12, "Sharing with Facebook and Other Social Networks," introduces you to the fascinating world of social networking—and shows you how to share with friends on Facebook, Pinterest, and Twitter.

Chapter 13, "Watching TV and Movies Online," shows you how to use your computer to watch television programming, movies, and other videos from Netflix, Hulu, and YouTube.

Chapter 14, "Playing Digital Music," shows you how to stream your favorite music online from Spotify and Pandora, as well as how to purchase and download music from Apple's iTunes Store—and listen to CDs on your computer with the Windows Media Player app.

Chapter 15, "Viewing and Editing Digital Photos," helps you connect a digital camera to your PC, transfer your photos to your PC, touch up problem pictures, and view them on your computer screen.

Chapter 16, "Protecting Your Computer," is all about defending against online menaces, keeping your PC running smoothly, backing up your important files, and recovering from serious crashes.

And that's not all. At the back of the book, you'll find a glossary of common computer terms—so you can understand what all the techie types are talking about!

(By the way, if something looks a little different on your computer screen than it does in your book, don't dismay. Microsoft is constantly doing little updates and fixes to Windows, so it's possible the looks of some things might have changed a bit between my writing this book and your reading it. Nothing to worry about.)

So, is using a computer really this easy? You bet—just follow the simple step-by-step instructions, and you'll be computing like a pro!

UNDERSTANDING PERSONAL COMPUTERS

Chances are you're reading this book because you have a new computer. At this point, you might not be totally sure what it is you've gotten yourself into. Just what is this mess of boxes and cables—how does it all go together, and how does it work?

We start by looking at the physical components of your system—the stuff we call computer *hardware*. A lot of different pieces and parts make up a typical computer system, and the pieces and parts differ depending on the type of computer you have.

You see, no two computer systems are identical. That's because there are several types of configurations (desktops, notebooks, and such) and because you can always add new components to your system—or disconnect other pieces you don't have any use for.

TYPES OF COMPUTERS

All-in-one
desktop PC

Traditional
desktop PC

Tablet PC

Notebook PC

GETTING TO KNOW DESKTOP PCS

A traditional desktop computer is one with a monitor designed to sit on your desktop, along with a separate keyboard and mouse and freestanding stereo speakers. The central component of a traditional desktop system is the *system unit*, which contains the PC's central processing unit (CPU), memory, and motherboard. All the external components connect directly to the system unit.

Start

System unit

Monitor

Keyboard

Mouse

End

NOTE

Connecting Components The external components (called *peripherals*) of a desktop PC connect to the system unit via an assortment of connectors. Most peripherals today connect via USB connectors, but some components use other types of connections. ■

NOTE

Wired and Wireless Connections On a desktop PC, most of the primary components connect to ports found on the back (or sometimes the front) of the system unit. However, some peripherals connect wirelessly, usually via Bluetooth. ■

GETTING TO KNOW ALL-IN-ONE PCS

An all-in-one computer is a desktop model in which the system unit is built in to the monitor. The monitor/system unit also includes built-in speakers, as well as all the ports you need to connect external peripherals. Many people like the easier setup and smaller space requirements of an all-in-one system.

Start

Monitor/
system unit

Mouse

Keyboard

End

NOTE

Touchscreens Some all-in-one PCs feature touch-screen monitors; you can control them by tapping and swiping the monitor screen with your fingers. ■

CAUTION

All-in-One Drawbacks The chief drawbacks to all-in-one systems are the price (usually a bit more than traditional desktop PCs) and the fact that if one internal component goes bad, the whole system is out of commission. It's a lot easier to replace a single component than an entire system! ■

GETTING TO KNOW NOTEBOOK PCS

Most new computers today are notebook models—sometimes called *laptops*. A notebook PC differs from a desktop PC in that all the pieces and parts are combined into a single unit that you can take with you almost anywhere. The built-in battery provides power when you're not near a wall outlet. And some notebook PCs include touchscreen displays, which let you operate Windows with a swipe of your fingertips.

Start

Display

Touchpad

Keyboard

End

NOTE

Types of Notebooks There are three types of notebook computers. *Traditional notebooks* have screens in the 14-inch to 16-inch range, 500GB or larger hard drives, and, in many cases, built-in CD/DVD drives. *Desktop replacement notebooks* have larger 17-inch screens and more powerful processors, but shorter battery life. *Ultrabooks* have smaller screens in the 10-inch to 14-inch range, no CD/DVD drive, but much longer battery life. Many ultrabooks also use faster solid-state memory rather than hard drives for storage. ■

TIP

External Peripherals Even though a notebook PC has the keyboard, mouse, and monitor built in, you can still connect external keyboards, mice, and monitors to the unit. This is convenient if you want to use a bigger keyboard or monitor or a real mouse (instead of the notebook's touchpad). ■

GETTING TO KNOW TABLET PCS

A tablet PC is a self-contained computer you can hold in one hand. Think of a tablet as the real-world equivalent of one of those communication pads you've seen on *Star Trek*. It doesn't have a separate keyboard; instead, you operate it by tapping and swiping the screen with your fingers. If you have a tablet that runs the Windows 10 operating system, you see a special tiled touch-centric interface. Some tablets come with optional keyboards and mice for office use.

Start

———— Touchscreen display

———— Power button

External ports

End

NOTE

Popular Tablets The most popular tablet PC today is the Apple iPad—which doesn't run Windows. (It runs Apple's own portable operating system, dubbed iOS.) There are numerous Windows-based tablets, however, including Microsoft's Surface tablets. ■

NOTE

Convertible PCs Several manufacturers offer *convertible* or *hybrid PCs*. A convertible PC is a blend of the ultrabook and tablet form factors; think of a convertible PC as an ultrabook that converts into a tablet or as a tablet that converts into an ultrabook. For example, the Asus Transformer Pad looks like an ultrabook but features a screen that detaches from the keyboard—which then functions as a freestanding touchscreen tablet. ■

CONNECTORS

Every external component you plug into your computer has its own connector, and not all connectors are the same. This results in an assortment of jacks—called *ports* in the computer world. The USB port is probably the most common, used to connect all sorts of external peripherals, including printers, keyboards, mice, and disk drives.

Start

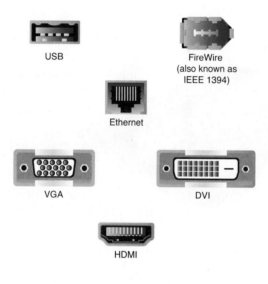

USB

FireWire
(also known as
IEEE 1394)

Ethernet

VGA

DVI

HDMI

End

NOTE

Portable Devices Most portable devices that you connect to your computer, such as smartphones and digital cameras, connect via USB—as do most larger peripherals. ■

TIP

HDMI If you want to connect your computer to your TV to watch Internet videos on the TV screen, look for a computer with an HDMI port. HDMI carries digital audio and high-definition video in a single cable. Most of today's flat-screen TVs have multiple HDMI inputs. ■

HARD DISK DRIVES: LONG-TERM STORAGE

The hard disk drive inside your computer stores all your important data—up to 6 terabytes (TB) or more, depending on your computer. A hard disk consists of metallic platters that store data magnetically. Special read/write heads realign magnetic particles on the platters, much like a recording head records data onto magnetic recording tape.

Start

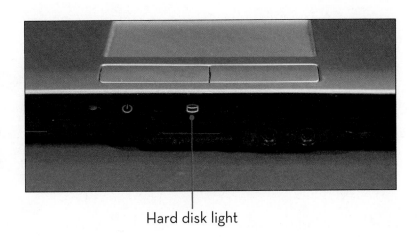

Hard disk light

End

TIP

Formatting the Drive Before you can store data on a hard disk, you must *format* the disk. When you format a hard disk, your computer prepares each track and sector of the disk to accept and store data magnetically. (Most new hard disks, such as the one in your new PC, come preformatted.) ■

NOTE

Ultrabook Storage Many ultrabook PCs use solid-state flash storage rather than hard disks. Solid-state storage is lighter and faster than hard disk storage—but it's more expensive and has a smaller storage capacity. ■

KEYBOARDS

A computer keyboard looks and functions just like a typewriter keyboard, except that computer keyboards have a few more keys (for navigation and special program functions). When you press a key on your keyboard, it sends an electronic signal to your system unit that tells your machine what you want it to do.

Start

Function keys

Alpha/
numeric keys

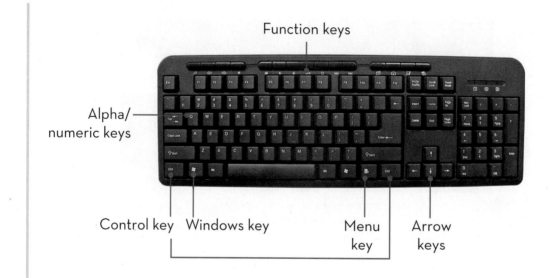

Control key   Windows key   Menu
key

Arrow
keys

End

NOTE

Windows Key Many essential operations are triggered by use of the special Windows key on the computer keyboard. (For example, you open the Windows Start menu by pressing the **Windows** key.) This key is indicated by the Windows logo. ■

TIP

Wireless Keyboards If you want to cut the cord, consider a wireless keyboard or mouse. These wireless devices operate via radio frequency signals and let you work several feet away from your computer, with no cables necessary. ■

TOUCHPADS

On a desktop PC, you control your computer's onscreen pointer (called a *cursor*) with an external device called a *mouse*. On a notebook PC, you use a small *touchpad* instead. Move your finger around the touchpad to move the cursor, and then click the left and right buttons below the touchpad to initiate actions in your program.

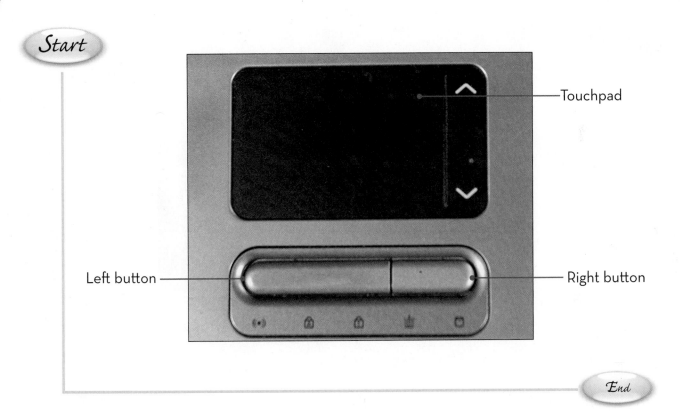

Start

Touchpad

Left button

Right button

End

TIP

External Mice If you'd rather use a mouse than a touchpad, you can connect any external mouse to your notebook PC via the USB port. Some manufacturers sell so-called notebook mice that are smaller and more portable than normal models. ■

NOTE

Mouse Options Most external mice offer more control options than built-in touchpads. For example, some mice include a *scrollwheel* you can use to quickly scroll through a web page or word processing document. ■

MEMORY CARD READERS

Many computers today include a set of memory card readers, usually grouped on the front or side of the unit. Memory cards store photos and movies recorded on digital cameras and camcorders. To read the contents of a memory card, simply insert the card into the proper slot of the memory card reader.

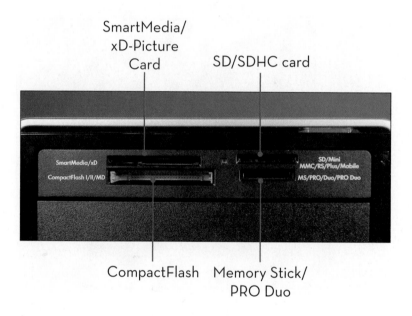

SmartMedia/
xD-Picture
Card

SD/SDHC card

CompactFlash

Memory Stick/
PRO Duo

NOTE

Memory Card Formats Different portable devices use different types of memory cards—which is why your computer has so many memory card slots. The most popular memory cards today are the Secure Digital (SD), Secure Digital High Capacity (SDHC), Secure Digital Extended Capacity (SDXC), and CompactFlash (CF) formats. ■

CD AND DVD DRIVES

Computer or data CDs, DVDs, and Blu-ray discs look just like the compact discs and movies you play on your home audio/video system. Data is encoded in microscopic pits below the disc's surface and is read from the disc via a drive that uses a consumer-grade laser. The laser beam follows the tracks of the disc and reads the pits, translating the data into a form your computer can understand.

Start

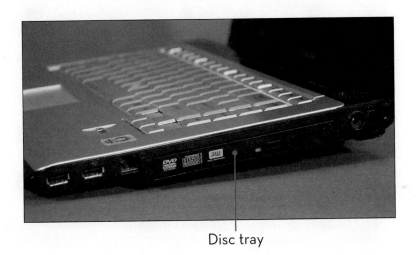

Disc tray

End

NOTE

CD, DVD, and Blu-ray Many new PCs come with combination CD/DVD drives that can read and write both CDs and DVDs. Some models include Blu-ray drives for high-definition video. But most ultrabooks and tablets don't come with a CD/DVD drive, helping to decrease weight and increase battery life. ■

NOTE

Music and Movies A computer CD drive can play back both data and commercial music CDs. A computer DVD drive can play back both data and commercial movie DVDs. ■

COMPUTER SCREENS

Your computer electronically transmits words and pictures to the computer screen built in to your notebook or to a separate video monitor on a desktop system. These images are created by a *video card* or chip installed inside the computer. Settings in Windows tell the video card or chip how to display the images you see on the screen.

Start

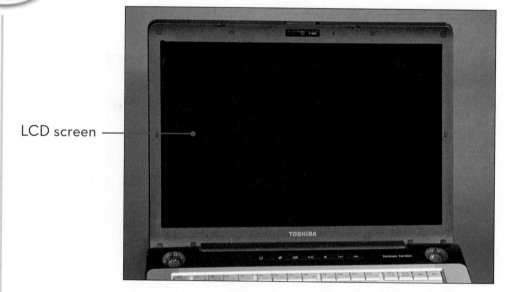

LCD screen —

End

NOTE

Touchscreen Displays Some notebook PCs and desktop monitors (as well as all tablets) feature touchscreen displays. These displays function just like traditional displays but are also touch sensitive, which means that you can control your system by tapping and swiping the screen with your fingers. ■

PRINTERS

To create a hard copy of your work, you must add a printer to your system. The two most common types are *laser* printers and *inkjet* printers. Laser printers work much like copy machines, applying toner (powdered ink) to paper by using a small laser. Inkjet printers shoot jets of ink onto the paper's surface to create the printed image.

Start

Operating buttons

Paper tray

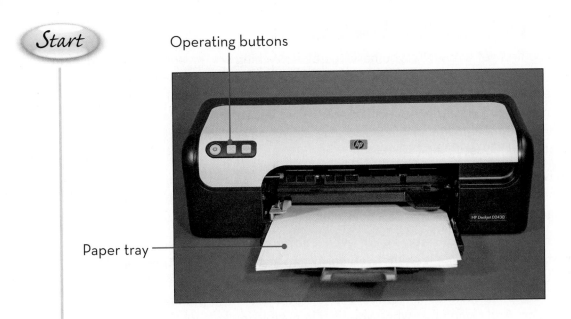

End

TIP

Black and White Versus Color Black-and-white printers are faster than color printers and better if you're printing memos, letters, and other single-color documents. Color printers are essential if you want to print pictures taken with a digital camera. ■

NOTE

Multifunction Printers So-called multifunction printers offer copy, scan, and fax functionality, in addition to traditional printing. ■

SETTING UP YOUR PC

When you first get a new PC, you have to get everything set up, connected, and ready to run. If you're using a traditional desktop PC, setup involves plugging in all the external devices—your monitor, speakers, keyboard, and such. If you're using an all-in-one desktop, the task is a bit easier because the system unit, monitor, and speakers are all in a single unit; all you have to connect are the keyboard and mouse.

Setup is even easier if you have a notebook PC, because all the major components are built in to the computer itself. Same thing with a tablet; there's really nothing to connect.

If you're connecting a desktop PC, or even a notebook with external peripherals, start by positioning it so that you easily can access all the connections on the unit. You'll need to carefully run the cables from each of the external peripherals to the main unit, without stretching the cables or pulling anything out of place. And remember, when you plug in a cable, make sure that it's *firmly* connected—both to the computer and to the specific piece of hardware. Loose cables can cause all sorts of weird problems, so be sure they're plugged in really well.

DESKTOP COMPUTER SYSTEM

Monitor

System unit

Mouse

Keyboard

SETTING UP A NOTEBOOK PC

Setting up a notebook PC is much simpler than setting up a desktop model. That's because almost everything is built in to the notebook—except external peripherals, such as a printer. Just connect the printer, plug your notebook into a power outlet, and you're ready to go.

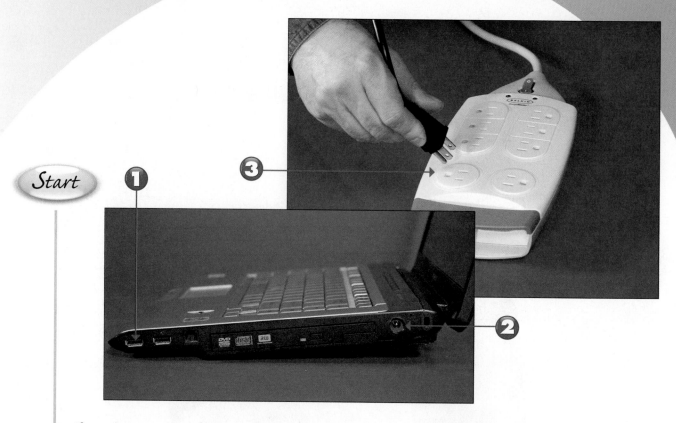

Start

③

①

②

End

① If you have a printer, connect one end of your printer's USB cable to a USB port on your notebook; connect the other end of the cable to your printer.

② Connect one end of your computer's power cable to the power connector on the side or back of your notebook; connect the other end of the power cable to a power source.

③ Connect your printer and other powered external peripherals to an appropriate power source.

TIP

External Peripherals If you're using an external mouse or keyboard, connect it to a USB port on your notebook. If you're using an external monitor, connect it to your notebook's external video port. ∎

SETTING UP AN ALL-IN-ONE DESKTOP PC

In an all-in-one desktop PC, the speakers and system unit are built in to the monitor, so you have fewer things to connect—just the mouse, keyboard, and any external peripherals, such as a printer. This makes for a quicker and easier setup than with a traditional desktop PC.

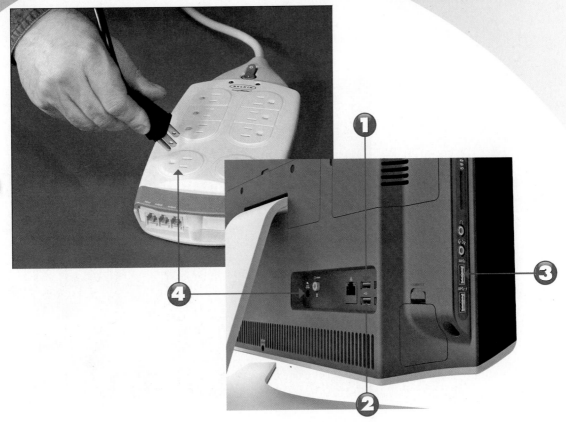

Start

1 Connect the mouse cable to a USB port on the back of the monitor.

2 Connect the keyboard cable to a USB port on the back of the monitor.

3 Connect one end of your printer's USB cable to a USB port on the back or side of your system unit; connect the other end of the cable to your printer.

4 Connect one end of your computer's power cable to the power connector on the back of your system unit; connect the other end of the power cable to a power source. Connect any powered peripherals to a power source.

End

TIP

Back and Side Connections Most all-in-one PCs have USB ports on both the back and the side of the unit. It doesn't matter which of these ports you use, although connecting to the back ports is usually a little cleaner looking—it does a better job of hiding the cables from view. ■

NOTE

External Speakers Some all-in-one PCs feature a speaker output you can use to add additional external speakers or perhaps a subwoofer (for better-sounding bass). On other all-in-ones, you can connect external USB speakers to an open USB port, if you like. ■

SETTING UP A TRADITIONAL DESKTOP PC

If you have a traditional desktop computer, you need to connect all the pieces and parts to your computer's system unit before powering it on. After connecting all your peripherals, you can then connect your system unit to a power source. Just make sure the power source is turned off before you connect!

1 Connect the mouse cable to a USB port on your system unit.

2 Connect the keyboard cable to a USB port on your system unit.

3 Connect the blue monitor cable to the blue monitor port on your system unit; make sure the other end is connected to your video monitor. (If your monitor has a DVI or HDMI connection, use that instead.)

Continued

NOTE

Mice and Keyboards Most newer mice and keyboards connect via USB. Some older models, however, connect to dedicated mouse and keyboard ports on your system unit. You should use whatever connection is appropriate. ■

TIP

Digital Connections Some newer computer monitors use a Digital Video Interface (DVI) or HDMI (High-Definition Multimedia Interface) connection instead of the older Video Graphics Array (VGA) type of connection. If you have a choice, a DVI or HDMI connection delivers a crisper picture than the older analog connection. HDMI is preferred if you're connecting to a flat-screen TV or home theater system because it transmits both video and audio. ■

4 Connect the green phono cable from your main external speaker to the audio-out or sound-out connector on your system unit; connect the other end of the cable to the speaker.

5 Connect one end of your printer's USB cable to a USB port on the back of your system unit; connect the other end of the cable to your printer.

Continued

TIP

Your Connection May Vary Not all speaker systems connect the same way. For example, some systems run the main cable to one speaker (such as the sub-woofer) and then connect that speaker to the other speakers in the systems. Other systems connect via USB. Make sure to read the manufacturer's instructions before you connect your speaker system. ■

NOTE

Connect by Color Most PC manufacturers color-code the cables and connectors to make connecting things even easier. Just plug the blue cable into the blue connector and so on. ■

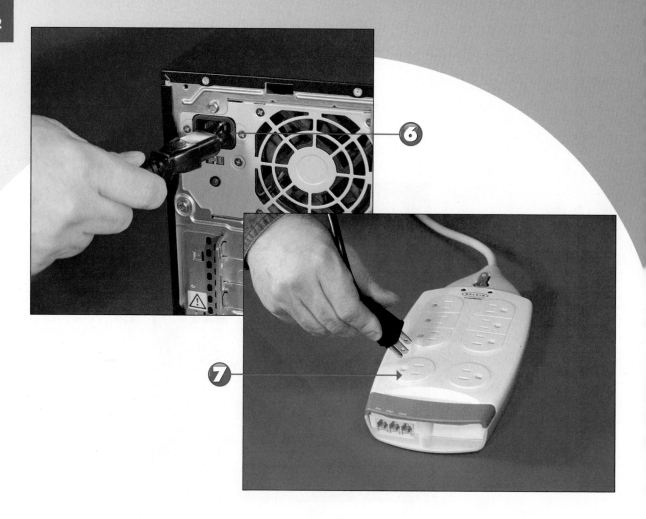

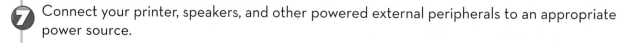

6 Connect one end of your computer's power cable to the power connector on the back of your system unit; connect the other end of the power cable to a power source.

7 Connect your printer, speakers, and other powered external peripherals to an appropriate power source.

End

TIP
Use a Surge Suppressor For extra protection, connect the power cable on your system unit to a surge suppressor rather than directly into an electrical outlet. This protects your PC from power-line surges that can damage its delicate internal parts. ■

CAUTION
Power Surges A power surge, whether from a lightning strike or due to an issue with your electric company, can do significant damage to a computer system. Too much power, even for just a second, can destroy your computer's microprocessor, memory chips, and other delicate components. In many instances, recovery from a power surge is either costly or impossible. ■

POWERING ON

Now that you have everything connected, sit back and rest for a minute. Next up is the big step: turning it all on!

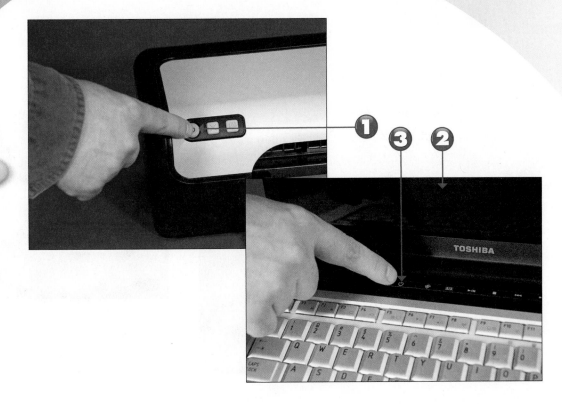

Start

End

① Turn on your printer, monitor (for a traditional desktop PC), and other powered external peripherals.

② If you're using a notebook PC, open the notebook's case so that you can see the screen and access the keyboard.

③ Press the power or "on" button on your computer.

NOTE

Booting Up Technical types call the procedure of starting a computer *booting* or *booting up* the system. Restarting a system (turning it off and then back on) is called *rebooting*. ■

CAUTION

Go in Order Your computer is the *last* thing you turn on in your system. That's because when it powers on it has to sense all the other components—which it can do only if the other components are plugged in and turned on. ■

LOGGING ON TO WINDOWS

Windows launches automatically as your computer starts. After you get past the Windows lock screen, you're taken directly to the Windows Start screen, and your system is ready to run.

Start

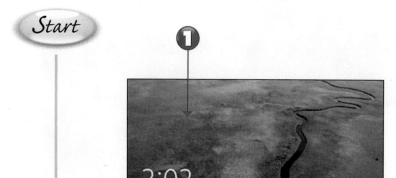

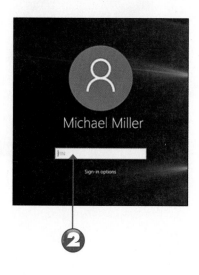

When you start your PC, you see the Windows lock screen; press any key to display your logon information.

Enter your password (if necessary) and press **Enter** on your keyboard.

End

TIP

Starting for the First Time The first time you start your new PC, you're asked to perform some basic setup operations, including activating and registering Windows and configuring your system for your personal use. ■

NOTE

Lock Screen Information The Windows lock screen displays a photographic background with some useful information on top—including the date and time, power status, and Wi-Fi (connectivity) status. ■

SHUTTING DOWN

When you want to turn off your computer, you do it through Windows. In fact, you don't want to turn off your computer any other way. You *always* want to turn things off through the official Windows procedure.

Start

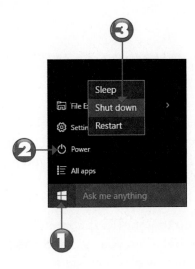

1. Click the **Start** button at the far left side of the taskbar to display the Start menu.

2. Click **Power** to display the submenu of options.

3. Select **Shut Down**.

End

TIP

Sleep Mode If you're using a notebook or tablet PC, Windows includes a special Sleep mode that keeps your computer running in a low-power state, ready to start quickly when you open the lid or turn it on again. In most instances, you enter Sleep mode by closing the tablet cover or the lid of your notebook. ■

Chapter 3

CONNECTING PERIPHERALS AND OTHER DEVICES

If you just purchased a brand-new, right-out-of-the-box personal computer, it probably came equipped with all the components you could ever desire—or so you think. At some point in the future, however, you might want to expand your system—by adding a printer, a webcam, a USB hub, or something equally new and exciting.

Everything that's hooked up to your PC is connected via some type of *port*. A port is simply an interface between your PC and another device, either internal (inside your PC's system unit) or external (via a connector on the back of the system unit). Different types of hardware connect via different types of ports.

USB CONNECTOR

USB type A connector (found on most PCs)

USB type B connector (found on some peripherals)

CONNECTING DEVICES VIA USB

Most external devices—including printers and smartphones—connect to your PC via USB. This is a type of connection common on computers and other electronic devices; it carries data and provides power for some connected devices. USB is popular because it's so easy to use. All you have to do is connect a device via USB and your computer should automatically recognize it.

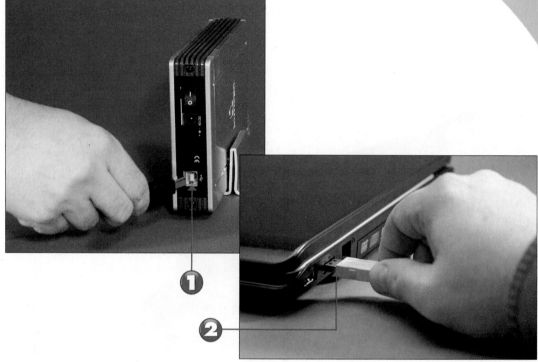

Start

 Connect one end of the USB cable to your new device.

2 Connect the other end of the cable to a free USB port on your PC.

End

NOTE

USB USB, which stands for *universal serial bus*, is an industry standard developed in the mid-1990s. There have been multiple versions of USB to date. All USB cables use similar connectors, but each successive version transmits data faster than previous versions. The most common version is USB 3.0, although USB 2.0 ports and cables are still common. The latest, USB-C, is just beginning to show up in new hardware. ■

TIP

USB Hubs If you connect too many USB devices, you can run out of USB connectors on your PC. If that happens, buy an add-on USB hub, which lets you plug multiple USB peripherals into a single USB port. ■

CONNECTING A PRINTER

Most printers today connect to your computer via an easy-to-use USB cable. After you've connected the printer, you can then configure it from within Windows.

Start

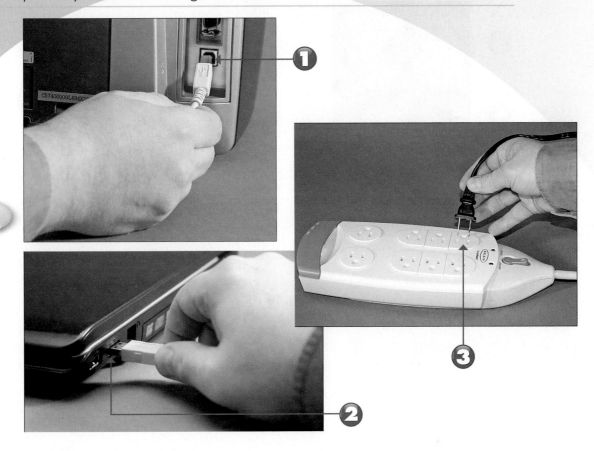

① Connect one end of a USB cable to the USB port on your printer.

② Connect the other end of the USB cable to a USB port on your system unit.

③ Connect the printer's power cable to a power outlet.

Continued

NOTE

Older Printers Many older printers used a different type of connection cable, called a *parallel cable*. Newer printers use USB connections because they can be connected without powering down, as was required with parallel connections. (In addition, USB cables are a lot thinner than parallel cables, which saves on space.) ■

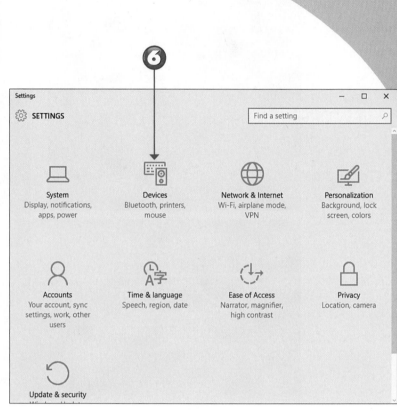

4 You must now install the printer within Windows. Click the **Start** button to display the Start menu.

5 Click **Settings** to display the Settings window.

6 Click **Devices**.

Continued

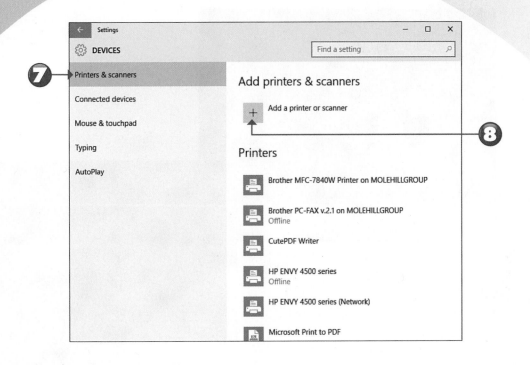

7 In the left-hand column, make sure **Printers & Scanners** is selected.

8 Scroll to the Printers section. If your printer is not listed, click the **Add a Printer or Scanner** button and let Windows search for it.

 End

CONNECTING YOUR PC TO YOUR TV

If you want to watch Internet streaming video (from Netflix and other services) on your TV, you can simply connect your TV to your personal computer via HDMI cable. Connected this way, anything you watch on your PC will display on your TV screen.

Start

 Connect one end of an HDMI cable to the HDMI port on your computer.

 Connect the other end of the HDMI cable to an open HDMI connector on your TV.

Continued

NOTE

HDMI HDMI, which stands for *High-Definition Multimedia Interface*, has become the connection standard for high-definition TVs. Most TV sets today have two or more HDMI inputs, typically used to connect cable boxes, Blu-ray players, and the like. HDMI transmits both audio and video signals. ■

NOTE

Mini HDMI Connectors Not all PCs have HDMI ports. Some have mini HDMI connectors, which require the use of a special HDMI cable with a mini plug on one end and a standard plug on the other. ■

Input Selection
AV1
AV2/S
AV3
COMPONENT
PC
HDMI1
HDMI2 [BLU-RAY]
HDMI3
HDMI4
TV
Media Server

Select
Change — EXIT
— RETURN

③

④

③ Switch your TV to the HDMI input you connected to. Your computer screen should now appear on your TV display.

④ To view programming full-screen, click the full-screen button in the app or window you're viewing.

End

Chapter 4

SETTING UP A WIRELESS HOME NETWORK

When you want to connect two or more computers in your home, you need to create a computer *network*. A network is all about sharing; you can use your network to share files, peripherals (such as printers), and even a broadband Internet connection.

There are two ways to connect your network: wired and wireless. A wireless network is more convenient (no wires to run), which makes it the network of choice for most home users. Wireless networks use radio frequency (RF) signals to connect one computer to another. The most popular type of wireless network uses the Wi-Fi standard and can transfer data at 11Mbps (802.11b), 54Mbps (802.11g), 600Mbps (802.11n), or 1Gbps (802.11ac).

UNDERSTANDING HOW WIRELESS NETWORKS WORK

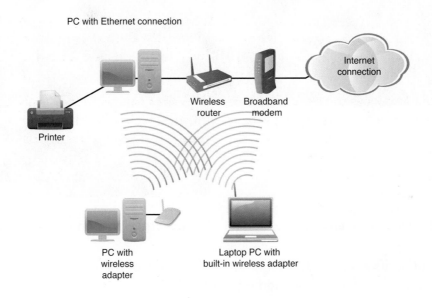

PC with Ethernet connection

Wireless router

Broadband modem

Internet connection

Printer

PC with wireless adapter

Laptop PC with built-in wireless adapter

SETTING UP YOUR NETWORK'S MAIN PC

The focal point of your wireless network is the *wireless router*. The wireless PCs on your network must be connected to or contain *wireless adapters*, which function as mini-transmitters/receivers to communicate with the base station.

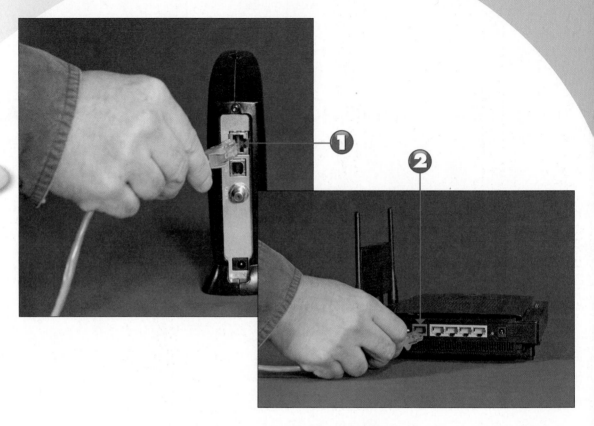

Start

Connect one end of an Ethernet cable to the Ethernet port on your broadband modem.

Connect the other end of the Ethernet cable to one of the Ethernet ports on your wireless router—preferably the one labeled Internet or WAN.

Continued

TIP

Internet Port Most routers have a dedicated input for your broadband modem, sometimes labeled Internet—although the modem can be connected to any open Ethernet input on the router. ■

NOTE

Internet Gateway Some Internet service providers (ISPs) provide broadband modems that include built-in wireless routers, often called an *Internet gateway*. If you have one of these devices, you don't need to buy a separate router. ■

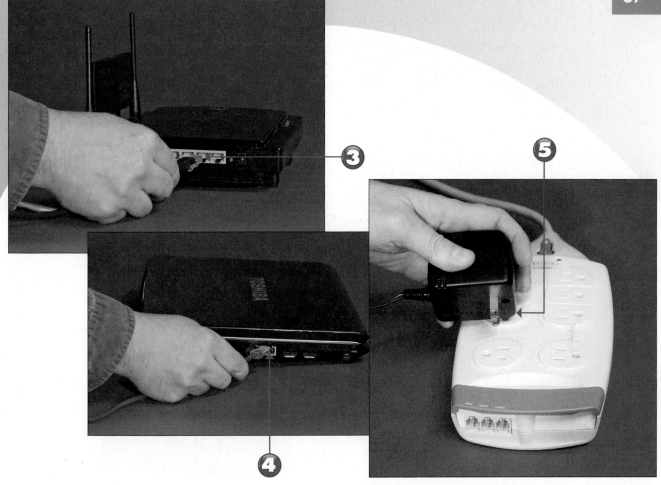

 Connect one end of an Ethernet cable to another Ethernet port on your wireless router.

 Connect the other end of the Ethernet cable to the Ethernet port on your main PC.

5 Connect your wireless router to a power source and, if it has a power switch, turn it on. Your computer should now be connected to the router and your network.

End

TIP

Router Configuration Some wireless routers require you to connect your main computer via Ethernet for initial configuration, as described here. Other routers will connect wirelessly to your main computer for the entire configuration process. When in doubt, follow the instructions that came with your router. ■

TIP

Wireless Security To keep outsiders from tapping into your wireless network, you need to enable wireless security for the network. This adds an encryption key to your wireless connection; no other computer can access your network without this key. ■

CONNECTING ADDITIONAL PCS TO YOUR WIRELESS NETWORK

Each additional PC on your network requires its own wireless adapter. Most notebook PCs come with a wireless adapter built in. Some desktop PCs come with built-in wireless adapters; others might require you to connect an external adapter.

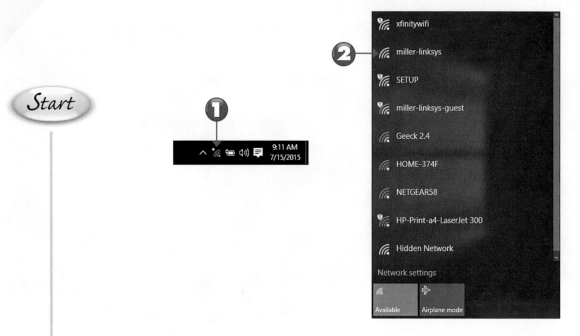

In the notification area of the taskbar, click the **Connections** icon to display the Settings window with the Wi-Fi panel displayed.

Click your wireless network; this expands the panel for this network.

Continued

NOTE

Connections Icon If no network is currently connected, the Connections icon should be labeled Not Connected—Connections Are Available. ■

TIP

Wireless Adapters A wireless adapter can be a small external device that connects to the PC via USB, an expansion card that installs inside your system unit, or a PC card that inserts into a laptop PC's card slot. ■

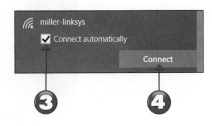

3 To connect automatically to this network in the future, check the **Connect Automatically** box.

4 Click **Connect**.

Continued

 TIP

Connect Automatically When you're connecting to your home network, it's a good idea to enable the Connect Automatically feature. This lets your computer connect to your network without additional prompting or interaction on your part. ∎

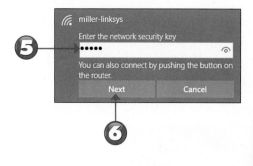

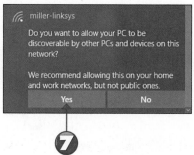

5 When prompted, enter the password (called the *network security key*) for your network.

6 Click **Next**.

7 When the next screen appears, click **Yes** to connect with other PCs and devices on your home network. You're now connected to your wireless router and should have access to the Internet.

End

TIP
Connecting Securely If you've enabled wireless security on your wireless router, you will be prompted to enter the passphrase or security key assigned during the router setup, as noted in steps 5 and 6. If you haven't enabled wireless security, you should. ■

TIP
One-Button Connect If your router supports "one-button wireless setup" (based on the Wi-Fi Protected Setup technology), you'll be prompted to press the "connect" button on the router to connect. You can connect via this button or by entering the network password as normal. ■

The easiest way to connect multiple home computers is to create a homegroup for your network. A *homegroup* is kind of a simplified network that lets you automatically share files and printers between connected computers.

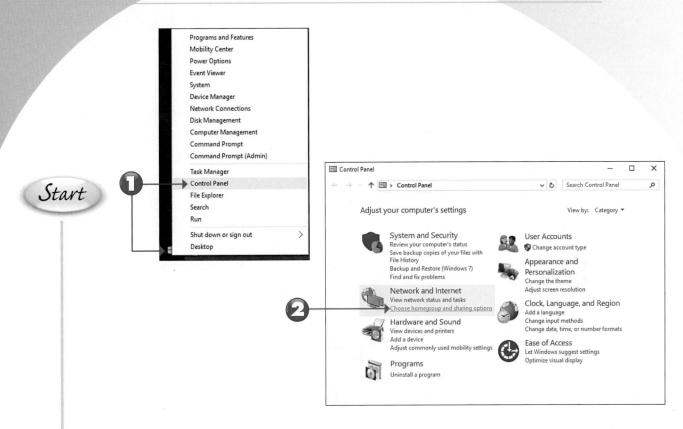

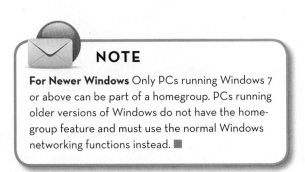

① Right-click the **Start** button to display the quick Access Menu, and then click **Control Panel**.

② In the Network and Internet section, click **Choose Homegroup and Sharing Options** to display the Share with Other Home Computers page.

Continued

NOTE

For Newer Windows Only PCs running Windows 7 or above can be part of a homegroup. PCs running older versions of Windows do not have the homegroup feature and must use the normal Windows networking functions instead. ■

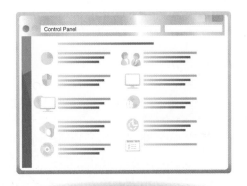

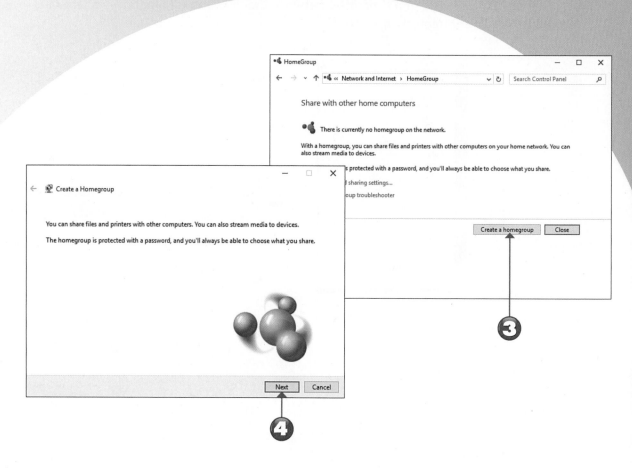

3 Click the **Create a Homegroup** button to display the Create a Homegroup page.

4 Click **Next** to display the Share with Other Homegroup Members page.

Continued

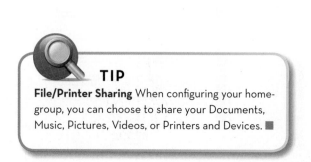

TIP

File/Printer Sharing When configuring your homegroup, you can choose to share your Documents, Music, Pictures, Videos, or Printers and Devices. ■

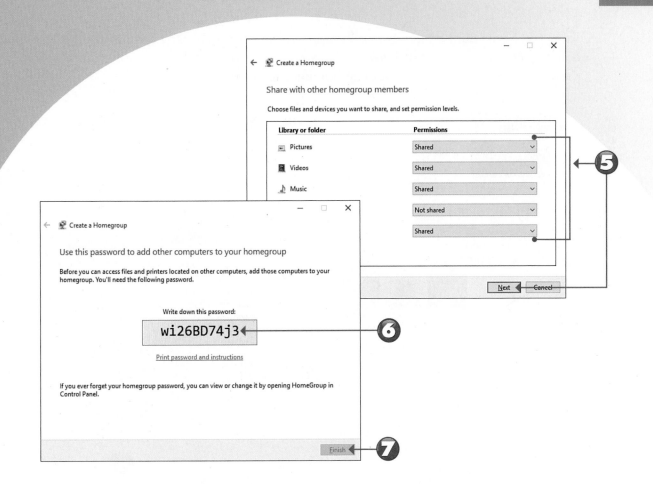

By default, Windows will share your pictures, videos, music, and printers and other devices—but not your documents—with other homegroup members. Make a selection in the drop-down list for each item to change the sharing permissions, then click **Next**.

Windows now displays the password for your new homegroup. You'll need to provide this to users of other computers on your network who want to join your homegroup, so write it down and keep it in a safe place.

Click the **Finish** button.

End

NOTE

Configuring Other PCs You'll need to configure each computer on your network to join your new homegroup. Enter the original homegroup password as instructed. ■

ACCESSING OTHER COMPUTERS IN YOUR HOMEGROUP

After you have your home network set up, you can access shared content stored on other computers on your network. How you do so depends on whether the other computer is part of your homegroup. We'll look at homegroup access first.

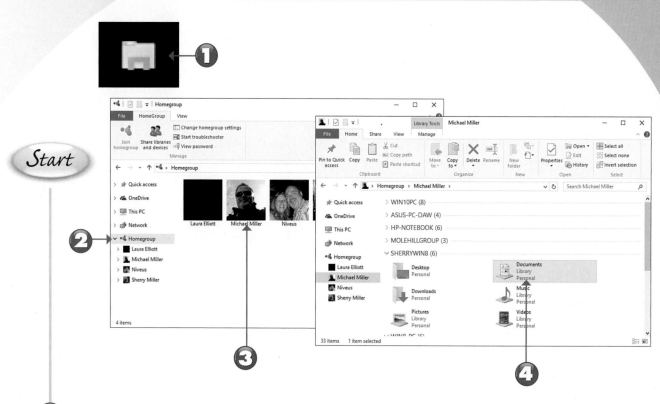

Start

1 Click the **File Explorer** icon on the Windows taskbar or on the Start menu.

2 When File Explorer opens, click **Homegroup** in the navigation pane to display all the users in your homegroup.

3 Double-click the user whose files you want to access.

4 Windows now displays the folders shared by that user. Double-click a folder to access that particular content.

End

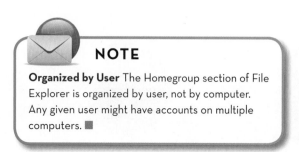

NOTE

Organized by User The Homegroup section of File Explorer is organized by user, not by computer. Any given user might have accounts on multiple computers. ∎

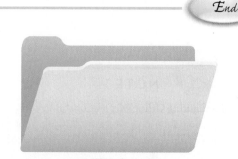

A computer doesn't have to be connected to your homegroup for you to access its content. Windows lets you access any computer connected to your home network—although you can share only content that the computer's owner has configured as sharable.

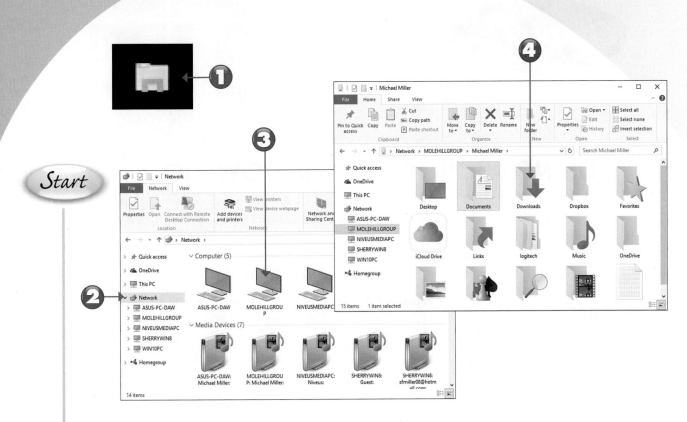

1. Click the **File Explorer** icon on the Windows taskbar or on the Start menu.

2. When File Explorer opens, click **Network** in the navigation pane. This displays all the computers and devices connected to your network.

3. Double-click the computer you want to access.

4. Windows now displays the shared folders on the selected computer. Double-click a folder to view that folder's content.

TIP
Look for the Public Folder On most older computers, shared files are stored in the Public folder. Look in this folder first for the files you want. ■

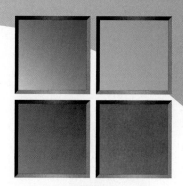

USING MICROSOFT WINDOWS 10

Microsoft Windows is a piece of software called an *operating system*. An operating system does what its name implies—it operates your computer system, working in the background every time you turn on your PC. The *desktop* that fills your screen is part of Windows, as is the taskbar at the bottom of the screen and the big menu that pops up when you click the Start button.

Windows 10 is the latest version of the Microsoft Windows operating system, the successor to Windows 8/8.1. If you used Windows 8 or 8.1, you'll appreciate the return to the traditional desktop and Start menu of the popular Windows 7.

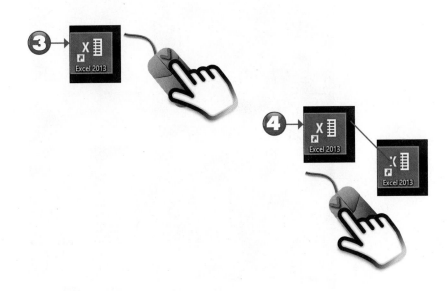

③ To right-click, position the cursor over the onscreen item and then click the *right* mouse button.

④ To drag and drop an item from one location to another, position the cursor over the item, click and hold the left mouse button, drag the item to a new position, and then release the mouse button.

End

TIP

Pop-Up Menus Many items in Windows feature a context-sensitive pop-up menu or Jump List. You access this menu or list by right-clicking the item. (When in doubt, right-click the item and see what pops up!) ■

TIP

Moving Files You can use dragging and dropping to move files from one folder to another or to delete files by dragging them onto the Recycle Bin icon. ■

USING THE WINDOWS START MENU

All the software programs and utilities on your computer are accessed via the Windows Start menu. Your most frequently used programs and basic Windows tools are listed on the left side of the Start menu; your favorite programs are "pinned" as tiles to the right side. To open a specific program, just click the icon or tile.

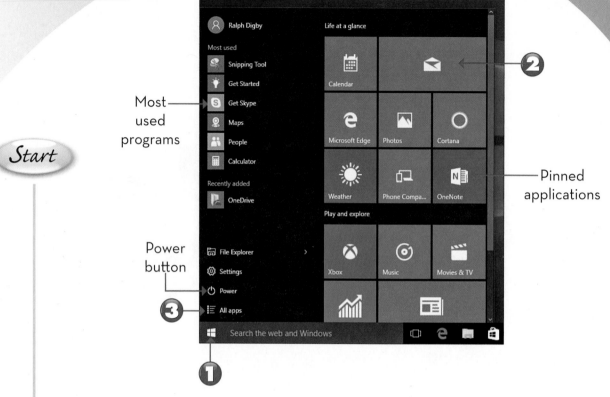

Most used programs

Start

Power button

Pinned applications

① Click the **Start** button to open the Start menu.

② Favorite programs are "pinned" to the right of the main Start menu in resizable tiles. Click a tile to open the application; scroll down to view more tiles.

③ Click **All Apps** to display a list of all installed applications.

Continued

NOTE

Windows 8/8.1 Start In Windows 8 and 8.1, Microsoft removed the Start menu, instead forcing users to use a new Start screen, designed for touchscreen use. In Windows 10 the Start menu is returned and the Start screen is removed. (Although tablet users can display the Start menu in full-screen mode.) ■

TIP

Shut Down To close Windows and shut down your computer, click the **Power** button on the Start menu and then select **Shut Down**. ■

4 Applications in the All Apps list are listed in alphabetical order, organized by letter; click an app to open it.

5 Some apps are organized in folders by publisher or type of application; click a folder to view its contents.

6 Click **Back** to return to the main Start menu.

End

TIP

Search for Apps You can search for apps installed on your computer by entering a program name into the Cortana Search box on the taskbar. ■

TIP

Quick Access Menu Right-click the Start button to display the Quick Access menu. This is a menu of advanced options, including direct links to File Explorer and Control Panel. ■

OPENING A PROGRAM

To open a program from the Start menu, all you have to do is click it. You can click an item in the applications list, or a tile on the right side of the menu.

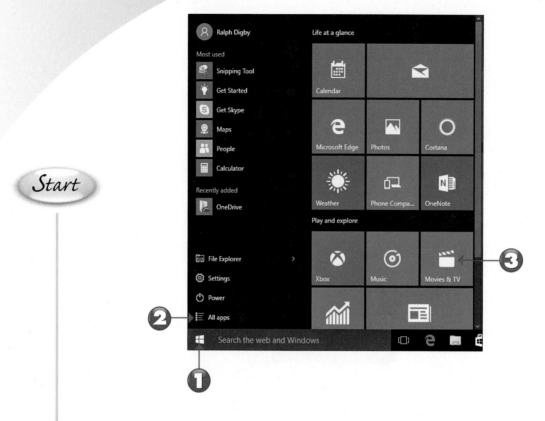

Start

1 Click the **Start** button to display the Start menu.

2 If the program you want isn't listed by default, click **All Apps** to display the complete list of applications.

3 Click the icon or tile for the program you want to launch.

End

USING THE TASKBAR

The taskbar is the area at the bottom of the Windows desktop. Icons on the taskbar can represent frequently used programs, open programs, or open documents.

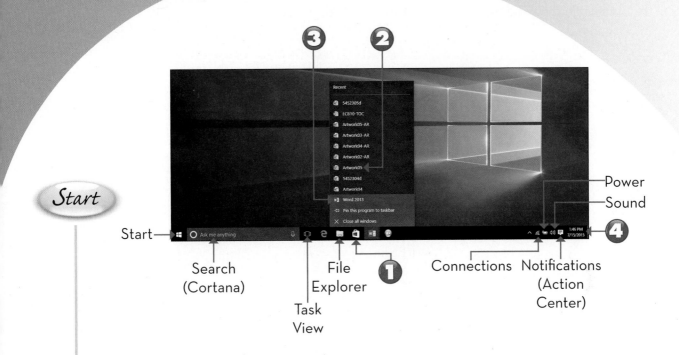

Start

Start — Start

Search (Cortana)

Task View

File Explorer

Connections

Notifications (Action Center)

Power

Sound

1 To open an application from the taskbar, click the application's shortcut icon.

2 To view a list of recently opened documents for an application, right-click the application's icon to display a Jump List, and then select an item.

3 To open a new blank document for an already-open application, right-click the application's icon to display the Jump List, and then click the application item.

4 The far right side of the taskbar is called the notification area, and it displays icons for essential Windows operations. To view more details about any item displayed in this area, click that item's icon.

End

TIP

Taskbar Icons A taskbar icon with a plain background represents an unopened application. A taskbar icon with a line underneath represents a running application. A taskbar icon with a shaded background represents the highlighted or topmost window on your desktop. An application with multiple documents open is represented by "stacked" lines under the icon. ■

TIP

Other Icons At the left side of the taskbar is a search box for Cortana, Windows 10's virtual personal search assistant. (Learn more about Cortana in Chapter 10, "Using the Internet.") There's also an icon for Task View, which enables you to create multiple virtual desktops with their own sets of open applications. In the notification area are icons for power (on notebook and tablet PCs), wireless connection status, sound, and notifications (opens the Action Center). ■

SCROLLING A WINDOW

Many windows contain more information than can be displayed in the window at once. When you have a long document or web page, only the first part of the document or page is displayed in the window. To view the rest of the document or page, you have to scroll down through the window, using the various parts of the scrollbar.

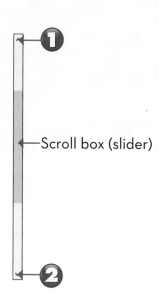

Scroll box (slider)

Start

1 Click the up arrow on the window's scrollbar to scroll up one line at a time.

2 Click the down arrow on the window's scrollbar to scroll down one line at a time.

End

TIP
Other Ways to Scroll To move to a specific place in a long document, use your mouse to grab the scroll box (also called a slider) and drag it to a new position. You can also click the scrollbar between the scroll box and the end arrow, which scrolls you one screen at a time. ■

After you've opened a window, you can maximize it to display full-screen. You can also minimize it so that it disappears from the desktop and resides as a button on the Windows taskbar, and you can close it completely.

Start

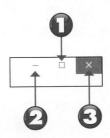

1 To maximize the window, click the **Maximize** button.

2 To minimize the window, click the **Minimize** button.

3 To close the window completely, click the **Close** (X) button.

End

TIP

Restoring a Window If a window is already maximized, the Maximize button changes to a Restore Down button. When you click the Restore Down button, the window resumes its previous (premaximized) dimensions. ■

SWITCHING BETWEEN OPEN WINDOWS

After you've launched a few programs, you can easily switch between one open program and another. In fact, Windows 10 offers several ways to switch programs.

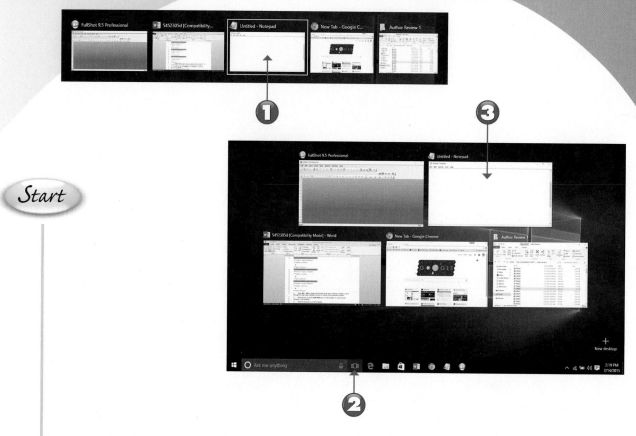

 Start

1. Press **Alt+Tab** to display thumbnails of all open windows; repeat to cycle through the open apps. Release the keys to switch to the selected window.

2. Alternatively, click the **Task View** icon on the taskbar to view all open windows in this desktop.

3. Click the window you want to display.

Continued

NOTE

Multiple Documents If multiple documents or pages for an application are open, multiple thumbnails will appear when you hover over that application's icon in the taskbar. ■

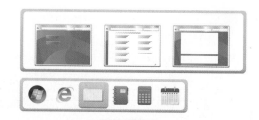

When a program or document is open, an icon for that item appears in the Windows taskbar. Mouse over that icon to view a thumbnail preview of all open documents for that application.

To switch to an open document from the taskbar, mouse over the icon for that item and then click the thumbnail for that document.

End

TIP

Help on the Web If you need technical help with using Windows, see Microsoft's website, www. microsoft.com. ■

USING WINDOWS IN TABLET MODE

If you have a notebook or desktop PC, Windows is displayed in the traditional desktop mode. If you have a tablet or hybrid PC, however, Windows can display in Tablet mode, specially designed for devices you operate via touch instead of a keyboard or mouse. Tablet mode is optimized for smaller screen devices, displaying the Start menu and all applications full-screen.

 Tap the **Start** button on the taskbar to display the Start menu full-screen.

Tap **All Apps** button to display the All Apps list.

 Tap any tile to launch the associated program in full-screen mode.

Continued

TIP

Manually Switch to Tablet Mode To manually switch to Tablet mode, click the **Notification** button on the taskbar to display the Action Center, and then click the **Tablet Mode** tile. ■

TIP

Continuum Automatic Switching The Windows Continuum feature automatically senses your device and displays in the correct mode. If you have a hybrid notebook/tablet device, when you remove the keyboard you will be asked whether you want to switch to tablet mode. When you reattach the keyboard, you'll be asked whether you want to switch to desktop mode. ■

4 To switch to other open apps, tap the Task View button on the taskbar.

5 To display the Action Center, swipe in from the right side of the screen or tap the Notifications icon in the taskbar.

End

USING WINDOWS WITH A TOUCHSCREEN DISPLAY

If you're using Windows on a computer or tablet with a touchscreen display, you use your fingers instead of a mouse to do what you need to do. So it's important to learn some essential touchscreen operations.

Start

1 On a touchscreen display, tapping is the equivalent of clicking with your mouse. Tap an item with the tip of your finger and release.

2 To display additional information about any item, press and hold the item with the tip of your finger.

Continued

TIP

Right-Click = Press and Hold Pressing and holding is the touchscreen equivalent of right-clicking an item with your mouse. ■

 To scroll down a page or perform many edge-centric operations, swipe the screen in the desired direction with your finger.

End

TIP

Zooming In To zoom in on a given screen (that is, to make a selection larger), use two fingers to touch two points on the item, and then move your fingers apart. ■

TIP

Zooming Out To zoom out of a given screen (that is, to make a selection smaller and see more of the surrounding page), use two fingers—or your thumb and first finger—to touch two points on the item, and then pinch your fingers in toward each other. ■

Chapter 6

PERSONALIZING WINDOWS

When you first turn on your new computer system, you see the Windows lock screen, and then the Windows desktop, complete with Start menu. If you like the way these items look, great. If not, you can change them.

Windows presents a lot of ways to personalize the look and feel of your system. In fact, one of the great things about Windows is how quickly you can make Windows look like *your* version of Windows, different from anybody else's.

SETTINGS WINDOW

Settings — window

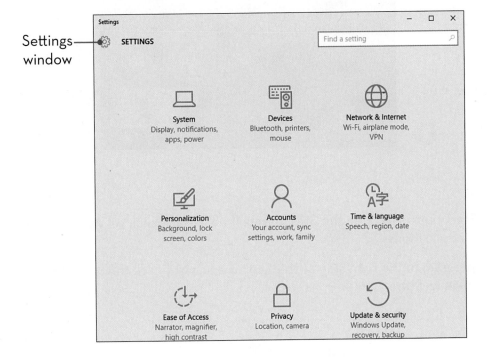

CUSTOMIZING THE START MENU

The Windows 10 Start menu can be customized to display tiles for your favorite programs. You can also resize the menu to take up more or less space on the desktop.

Start

② Pin to Start
   Pin to taskbar
   Don't show in this list

①

① Click the **Start** button to open the Start menu. If necessary, click **All Apps** to view a list of all installed applications.

② To "pin" a program to the right side of the Start menu, right-click the name of the app and then select **Pin to Start**.

Continued

NOTE

Pinning "Pinning" an app creates a permanent shortcut to that app. You can pin programs to either the Start menu or the taskbar. ▪

3 To rearrange tiles on the Start menu, click and hold a tile, and then drag it to a new position.

4 To resize a tile, right-click the tile, select **Resize**, and then select the desired size.

5 To remove a tile from the Start menu, right-click the tile and select **Unpin from Start**.

End

NOTE

Tile Sizes Tiles come in four possible sizes: Small, Medium, Wide, and Large. ■

NOTE

Live Tiles Tiles for some apps display "live" information—that is, current data in real time. For example, the Weather tile displays current weather conditions; the News tile displays current news headlines. ■

CHANGING THE DESKTOP BACKGROUND

One of the most popular ways to personalize the desktop is to use a favorite picture or color as the desktop background.

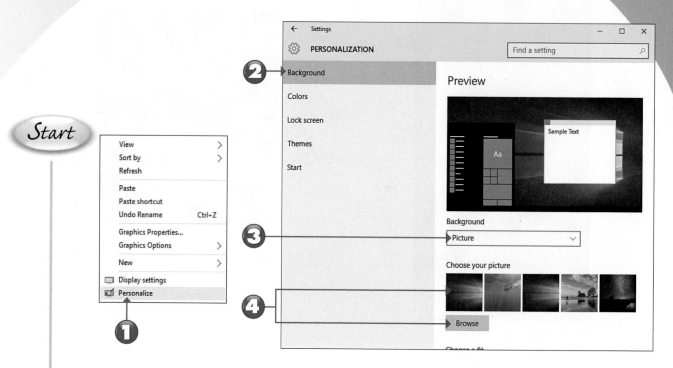

Start

1. Right-click any open area of the desktop to display the options menu, and then click **Personalize** to display the Personalization window.

2. Click to select the **Background** tab.

3. To use a picture background, click the **Background** list and select **Picture**.

4. Select one of the image thumbnails, or click **Browse** to select another picture stored on your computer.

Continued

TIP

Choose a Fit There are six options in the Choose a Fit list for displaying images that don't fit the entire desktop. **Fill** zooms into the picture to fill the screen. **Fit** fits the image to fill the screen horizontally—but might leave black bars above and below the image. **Stretch** distorts the picture to fill the screen. **Tile** displays multiple instances of a smaller image. **Center** displays a smaller image in the center of the screen, with black space around it. Select **Span** if you're using multiple monitors and want the same image to span the monitors. ▪

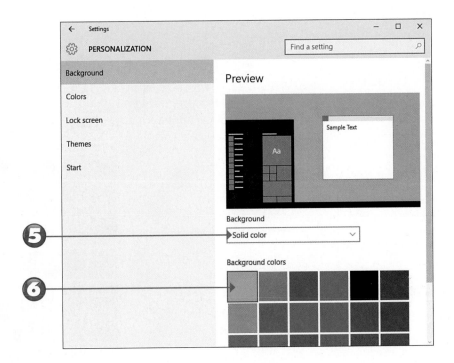

5 To select a solid-color background, click the **Background** list and select **Solid Color**.

6 Click the color you want for your desktop background.

End

TIP

Background Slide Show To display more than one image in a changing desktop slide show, click the **Background** list and select **Slideshow**. You can then click the **Browse** button to select which pictures appear in the slide show. Pull down the **Change Picture Every** list to determine how quickly images change. ■

CHANGING THE ACCENT COLOR

You can select any color for the title bar and frame that surround open windows on the desktop. You can also select a color for the background of the Start menu, taskbar, and Action Center pane.

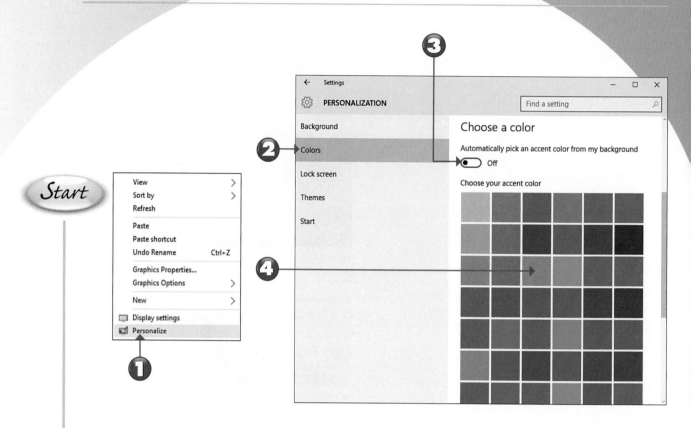

1 Right-click any open area of the desktop to display the options menu, and then click **Personalize** to display the Personalization window.

2 Click to select the **Colors** tab.

3 To select a specific color, click the **Automatically Pick an Accent Color from My Background** option to the Off position.

4 Select the color you want from the color chooser.

Continued

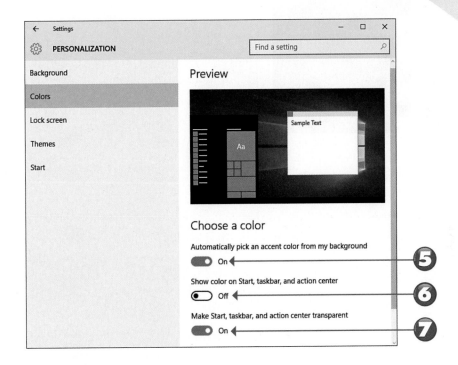

⑤ To have Windows automatically suggest a contrast color based on the color of the current background image, set the **Automatically Pick an Accent Color from My Background** option to the On position.

⑥ To have the Start menu, taskbar, and Action Center show the selected contrast color, click the **Show Color on Start, Taskbar, and Action Center** control to the On position. To have these items display in black, click this control to the Off position.

⑦ To apply a transparent effect to the Start menu, taskbar, and Action Center, click the **Make Start, Taskbar, and Action Center Transparent** control to the On position.

End

CUSTOMIZING THE LOCK SCREEN PICTURE

The lock screen is what you see when you first power on your computer or begin to log on to Windows. You can easily change the background picture of the lock screen to something you like better and add information from up to seven apps to the screen.

1 Click the **Start** button to display the Start menu.

2 Click **Settings** to display the Settings window.

3 Click **Personalization** to display the Personalization screen.

Continued

TIP

Lock Screen The lock screen appears when you first power on your PC and any time you log off from your personal account, switch users, or lock your computer. It also appears when you awaken your computer from Sleep mode. ■

NOTE

Smartphone Lock Screens The Windows lock screen is similar to the lock screens you see on various smartphones, such as the Apple iPhone, whenever you "wake up" the phone. ■

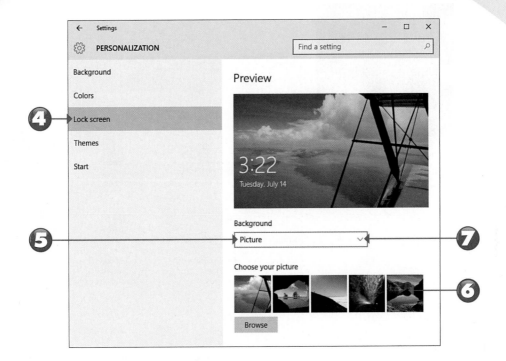

4 Click to select the **Lock Screen** tab.

5 Click the **Background** list and select **Picture**.

6 Click the thumbnail for the picture you want to use.

7 To display a slide show of pictures on your lock screen, click the **Background** list and select **Slideshow**.

End

TIP

Personalize Your Picture To use another picture as the lock screen background, click the **Browse** button. When the Open dialog box appears, navigate to and click or tap the picture you want to use, and then click the **Choose Picture** button. ■

ADDING APPS TO THE LOCK SCREEN

The lock screen can display a number of apps that run in the background and display useful or interesting information, even while your computer is locked. By default, you see the date/time, power status, and connection status, but it's easy to add other apps to the lock screen.

1 Click the **Start** button to display the Start menu.

2 Click **Settings** to display the Settings window.

3 Click **Personalization** to display the Personalization screen.

Continued

TIP

Real-Time Information The apps you see on the lock screen display information in real time. ■

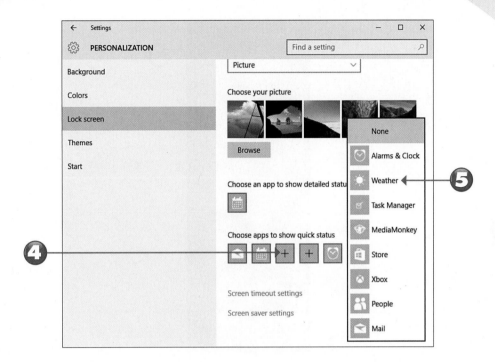

4 Go to the Choose Apps to Show Quick Status section and click a + button to display the list of options.

5 Click or tap the app you want to add.

End

TIP

Displaying Live Information You can also opt for one of the lock screen apps to display detailed live information, such as unread messages or current weather conditions. To select which app displays detailed information, click the button for that app in the Choose an App to Show Detailed Status section. ■

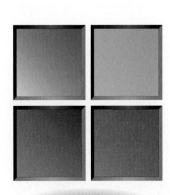

CHANGING YOUR PROFILE PICTURE

When you first configured Windows, you were prompted to select a default image to use as your profile picture. You can, at any time, change this picture to something more to your liking.

① Click the **Start** button to display the Start menu.

② Click your name or picture at the top of the Start menu to display the options menu.

③ Click **Change Account Settings** to display the Settings window with the Accounts page displayed.

Continued

4 Go to the Your Picture section and click one of the images displayed there, or...

5 Click the **Browse** button to open the Open window.

6 Navigate to and click or tap the picture you want.

7 Tap or click the **Choose Picture** button.

End

TIP
Webcam Picture If your computer has a webcam, you can take a picture with your webcam to use for your account picture. Scroll to the Create Your Picture section, click the **Camera** button, and follow the onscreen directions from there. ■

SETTING UP ADDITIONAL USERS

Chances are you're not the only person using your computer; it's likely that you'll be sharing your PC with your spouse and kids, at least to some degree. Fortunately, you can configure Windows so that different people in your household can sign on to the computer with their own custom settings—and have access to their own personal files. You do this by assigning each user in your household his own password-protected user account.

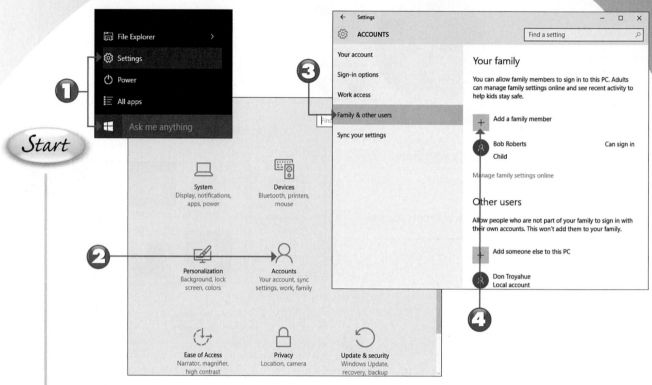

Start

① Click the **Start** button to display the Start menu, and then click **Settings** to display the Settings window.

② Click **Accounts** to display the Accounts page.

③ Click to select the **Family & Other Users** tab.

④ Click **Add a Family Member** to display the Add a Child or an Adult? window.

Continued

TIP

Two Types of Accounts Windows lets you create two types of user accounts: online and local. An *online account* is linked to a new or existing Microsoft account; a *local account* is exclusive to your current computer, and doesn't link to any online services. ■

NOTE

Microsoft Account By default, Windows creates new user accounts using existing or new Microsoft accounts. You need a Microsoft Account login to use many of the interactive features of Windows 10, such as linking your account to Facebook or Microsoft's OneDrive; a Microsoft account is also necessary to access features with live updates, such as the Weather and News apps. ■

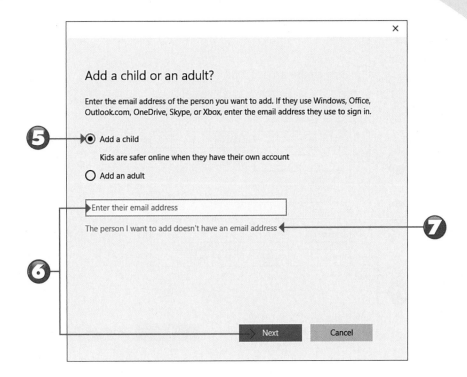

5 Select either **Add a Child** or **Add an Adult**.

6 If the person has an email address, enter it and click **Next**.

7 If this person doesn't have an email address, click **The Person I Want to Add Doesn't Have an Email Address**.

Continued

TIP

Three Ways to Log On When you set up an account, you can choose from three ways to log on. You can log on to an account with a traditional password, a PIN code, or a picture password. ■

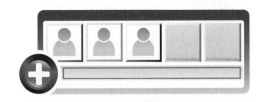

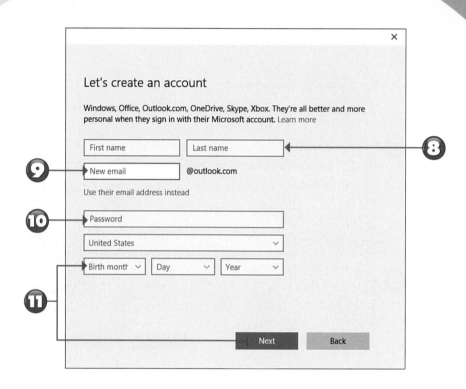

8 Enter the person's name into the **First Name** and **Last Name** boxes.

9 Enter the desired email username into the **New Email** box. (You might have to try several names to get one that isn't already taken.)

10 Enter the desired password into the **Password** box.

11 Enter this person's birthdate, and then click **Next** and follow the remaining instructions to create your new account.

End

TIP

Child Accounts If you're creating a child account, Windows automatically activates Family Safety Monitoring. With Family Safety Monitoring, you can turn on web filtering (to block access to undesirable websites), limit when the kids can use the PC and what websites they can visit, set limits on games and Windows Store app purchases, and monitor the youngsters' PC activity. ■

If other people are using your computer, they might want to log on with their own accounts. To do this, you'll need to change users—which you can do without shutting off your PC.

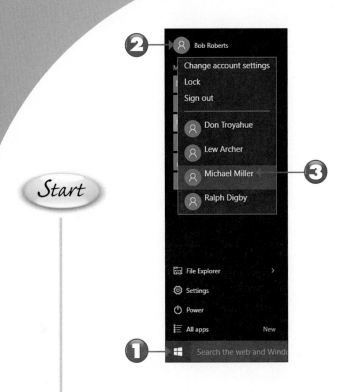

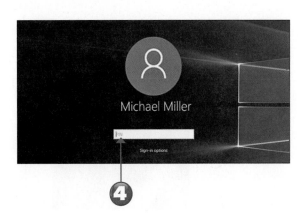

1 Click the **Start** button to display the Start menu.

2 Click your name or picture at the top of the Start menu.

3 Click the desired user's name.

4 When prompted, enter the new user's password, and then press **Enter**.

End

TIP

Signing Out When you switch users, both accounts remain active; the original user account is just suspended in the background. If you would rather log off completely from a given account and return to the Windows lock screen, click your username on the Start menu, and then click **Sign Out**. ■

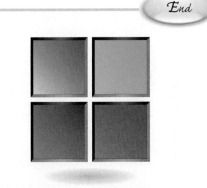

CONFIGURING WINDOWS SETTINGS

You can configure many other Windows system settings if you want. In most cases, the default settings work just fine and you don't need to change a thing. However, you can change these settings if you so desire. You configure most of these settings from the Settings window.

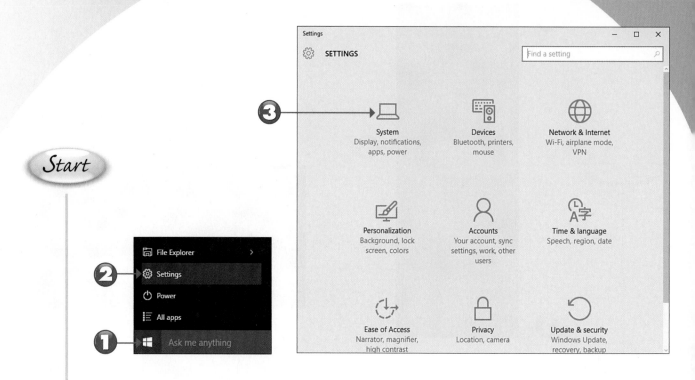

Start

1 Click the **Start** button to display the Start menu.

2 Select **Settings** to display the Settings window.

3 Select an option to display the associated settings.

Continued

TIP

Settings The following options are available in the PC Settings window: System, Devices, Network & Internet, Personalization, Accounts, Time & Language, Ease of Access, Privacy, Update & Recovery. ■

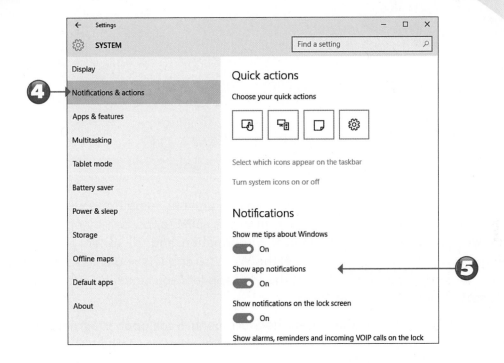

4 Click the setting you want to configure.

5 Configure the necessary options.

End

TIP
Control Panel You can also configure most system settings from the traditional Windows Control Panel. To open the Control Panel, right-click the lower-left corner of any screen to display the Quick Access menu, and then select **Control Panel**. ■

Chapter 7

WORKING WITH SOFTWARE APPLICATIONS

Most of the productive and fun things you do on your computer are done with *software programs* or *applications*, sometimes called *apps*. Some applications are work related, others provide useful information, and still others are more entertaining in nature. For example, the Weather app lets you check current weather conditions and forecasts; the Mail app lets you send and receive email messages over the Internet.

You open software programs from the Start menu. Each open program sits on the desktop in its own individual window. This enables you to have multiple open apps onscreen at the same time, with the windows stacked on top of or tiled next to each other.

PARTS OF A WINDOW

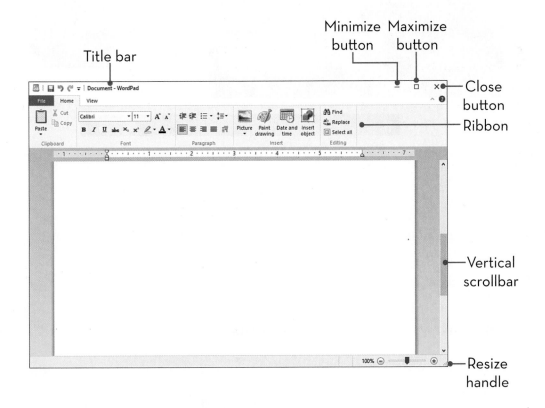

Title bar

Minimize button

Maximize button

Close button

Ribbon

Vertical scrollbar

Resize handle

You can open programs from the Start menu, the taskbar, or the desktop.

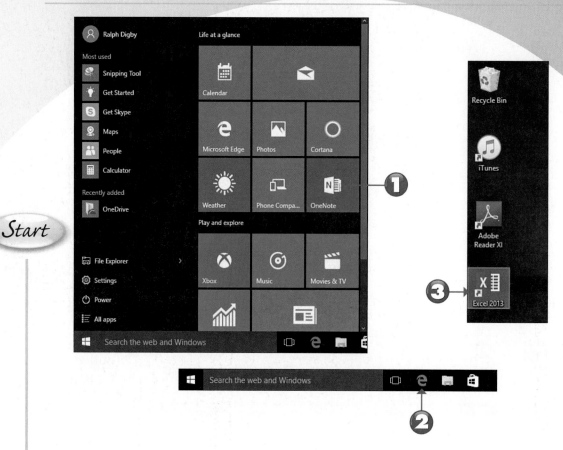

Start

① From the Start menu, click the name of or tile for the app you want to open.

② From the taskbar, click the icon for the app you want to open.

③ From the desktop, double-click the shortcut for the app you want to open.

End

You can "pin" your favorite applications to the Start menu. Programs you pin appear as resizable tiles on the right side of the Start menu.

1 Open the Start menu and navigate to the application you want to pin. You might need to click **All Apps** to view all your installed programs.

2 Right-click the name of the application to display the options menu.

3 Click **Pin to Start**. A tile for the app will now appear on the Start menu.

TIP
Moving Tiles To rearrange tiles on the Start menu, click and drag any tile to a new position. ■

TIP
Resizing Tiles To resize a tile on the Start menu, right-click the tile, select **Resize**, and then select a different size. ■

PINNING A PROGRAM TO THE TASKBAR

Instead of opening the Start menu whenever you want to launch a new program, you can instead "pin" shortcuts to your favorite programs to the desktop taskbar. You can then launch one of these programs by clicking the shortcut on the taskbar.

Open the Start menu and navigate to the application you want to pin. You might need to click **All Apps** to view all your installed programs.

Right-click the name of the application to display the options menu.

Click **Pin to Taskbar**. An icon for the app will now appear on the taskbar.

TIP
Rearranging Taskbar Icons To change the order of the apps you've pinned to the taskbar, use the mouse to click and drag an icon to a new position. ■

You can also add shortcuts to your favorite apps directly to the Windows desktop. These shortcuts appear as small icons on the desktop.

Start

End

1. Click the **Show Desktop** button at the far-right side of the taskbar to minimize all windows on the desktop.

2. Open the Start menu and navigate to the application for which you're creating a shortcut. You might need to click **All Apps** to view all your installed programs.

3. Click and drag the app from the Start menu onto the desktop. The menu item remains on the Start menu, but a shortcut to that item is placed on the desktop.

TIP

Organizing Desktop Shortcuts Use your mouse to click and drag shortcut icons to whatever position you want on the desktop. ■

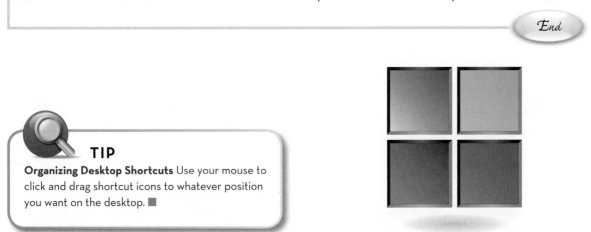

SEARCHING FOR APPS

If you have a lot of apps installed on your PC, finding the app you want, either on the Start menu or elsewhere, might be difficult. You can instead search for specific apps using Cortana, Windows 10's virtual personal assistant.

Start

1 Click within the **Cortana Search** (Ask Me Anything) box on the taskbar and start typing the name of the app you're looking for.

2 As you type, Cortana suggests apps that match your query in the Apps section of the Cortana panel. (Other types of matching items are also displayed.) Click the app you want to open.

End

TIP

Complete Your Query If Windows doesn't suggest the app you want, finish entering your query and then click or tap the magnifying glass button to start the search. You will then see a list of matching apps (and other items); click the app you want to open. ■

NOTE

Cortana When you search for an app, you're using Windows 10's Cortana feature. Cortana is a type of virtual assistant that simplifies searching both within and outside Windows, and offers additional task and time management features. Learn more about Cortana in Chapter 10, "Using the Internet." ■

In addition to traditional desktop apps, Microsoft offers what are called Universal Windows apps. These apps are subtly different from traditional software programs, and designed specifically for the Windows 10 experience.

Start

1 To review the basic settings of a Universal Windows app, click the **Options** button.

2 This opens an Options panel for that app. Click to access any app feature.

3 Click the **Settings** icon to configure additional settings.

End

NOTE

Universal The apps we call Universal or Windows apps used to be called Metro, Modern, and Windows Store apps. (Microsoft apparently can't make up its collective mind.) The word "universal" refers to the fact that these apps run on various devices, including personal computers, tablets, and smartphones. ■

NOTE

Evolving Universal Windows Apps Universal Windows apps, then called Metro apps, were first introduced in Windows 8, designed for touch-first operation on smaller touchscreen devices. In Windows 8 and 8.1, these apps ran full-screen, not on the desktop. With Windows 10, however, Microsoft has enabled these apps to run on the traditional desktop, in normal windows, and without the need for a touchscreen display. That means you can run these apps in their own windows and operate them with your computer's mouse and keyboard. ■

USING CONTEXT MENUS

Many onscreen elements have additional options that can be accessed via a "hidden" context-sensitive menu. You open this menu by right-clicking (instead of the normal left-clicking) the item.

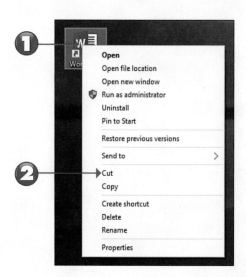

Open
Open file location
Open new window
Run as administrator
Uninstall
Pin to Start

Restore previous versions

Send to   >

Cut
Copy

Create shortcut
Delete
Rename

Properties

1 Right-click the item to display the context or options menu.

2 Click the action you want.

End

TIP

Keyboard Shortcut You can also display the context menu for an item by highlighting the item, with either your mouse or the keyboard arrow keys, and then pressing the Menu key on your keyboard. ■

USING PULL-DOWN MENUS

Many software programs use a set of pull-down menus to store all the commands and operations you can perform. The menus are aligned across the top of the window, just below the title bar, in what is called a *menu bar*. You open (or pull down) a menu by clicking the menu's name; you select a menu item by clicking it with your mouse.

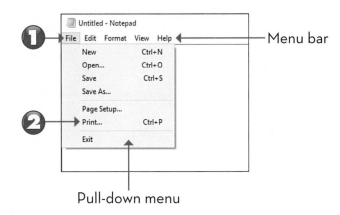

Menu bar

Pull-down menu

Start

1 Click the menu's name to pull down the menu.

2 Click the menu item to select it.

End

TIP

Not All Items Are Available If an item in a menu, toolbar, or dialog box is dimmed (or grayed), that means it isn't available for the current task. ■

USING TOOLBARS

Some software programs put the most frequently used operations on one or more *toolbars*, usually located just below the menu bar. A toolbar looks like a row of buttons, each with a small picture (called an *icon*) and maybe a bit of text. You activate the associated command or operation by clicking the button with your mouse.

Toolbar

1 Click a button on the toolbar to select that operation.

End

TIP

Long Toolbars If the toolbar is too long to display fully on your screen, you'll see a right arrow at the far-right side of the toolbar. Click this arrow to display the buttons that aren't currently visible. ■

USING RIBBONS

Some Windows programs use a *ribbon* interface that contains the most frequently used operations. A ribbon is typically located at the top of the window, under the title bar (and sometimes the menu bar). Ribbons often consist of multiple tabs; select a tab to see buttons and controls for related operations.

Start

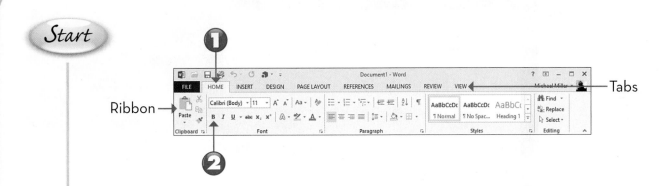

Ribbon

Tabs

1 Click a tab to select that particular set of functions.

2 Click a button on the ribbon to select that operation.

End

NOTE

Ribbons Versus Toolbars The ribbon interface is found in many newer applications. Most older applications use toolbars instead. ■

TIP

Display or Hide If the ribbon isn't visible, click the down arrow at the far-right side of the tabs. To hide the ribbon and its buttons, click the up arrow at the far-right side of the ribbon. ■

USING WINDOWS 10'S BUILT-IN APPS

Windows 10 ships with a number of useful apps built in to the operating system. Most of these apps are Universal apps that you can launch from the Start menu.

Start

① The Weather app displays current weather conditions as well as a multiday forecast and radar maps.

② The Maps app displays a map of your current location, as well as step-by-step directions to any location you want to visit.

Continued

TIP

Scroll for More Scroll down through the Weather app to view additional weather information, including an hourly forecast, various weather maps, and a graph for historical weather in your location. Click any item to view more detail. ■

NOTE

Bing Maps The Maps app is based on Bing Maps, which is Microsoft's web-based mapping service. ■

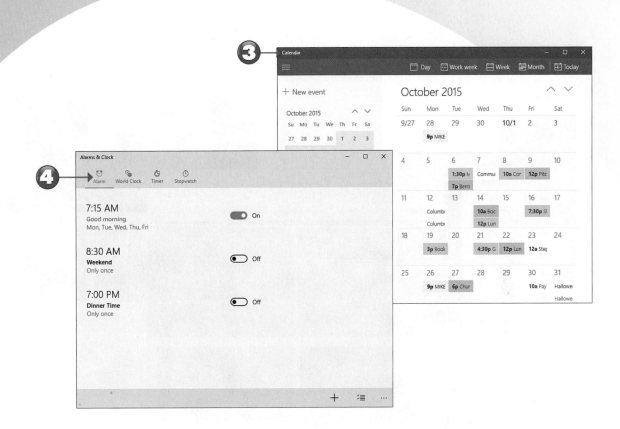

3 The Calendar app displays upcoming appointments in daily, weekly, or monthly views.

4 The Alarms & Clock app turns your computer into a digital alarm clock, and also includes timer and stopwatch functions.

Continued

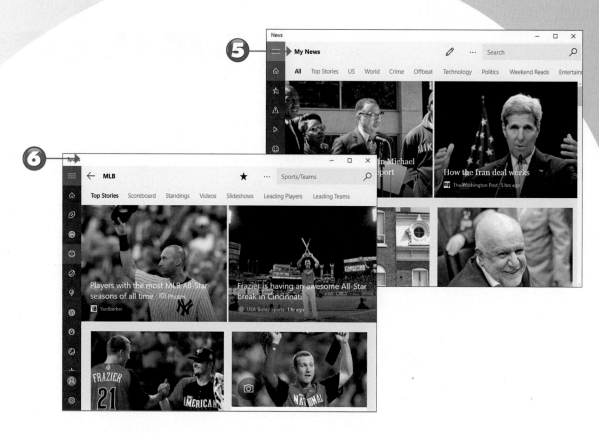

5 The News app displays the latest news headlines; click a headline or an image to read the full story.

6 The Sports app displays the latest sports headlines, as well as scores from your favorite teams.

Continued

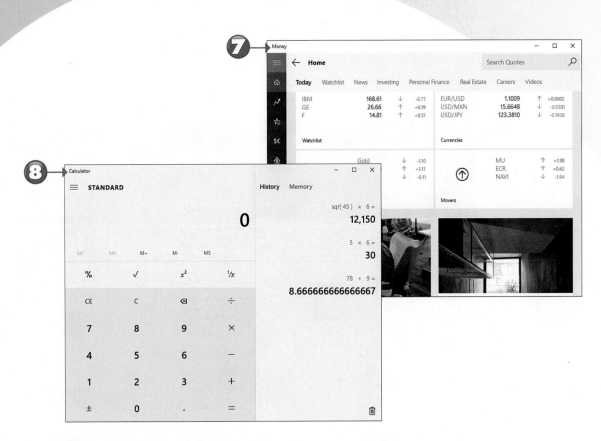

7 The Money app helps you stay up-to-date on the latest financial news and keep track of your personal investments.

8 Use the Calculator app as a standard, scientific, or programmer's calculator.

End

TIP

Watching Stocks The Money app lets you create a "watchlist" of stocks you own or want to track. To add a new stock to your watchlist, click the **+** tile. When the Add to Watchlist panel appears, enter the name or symbol of the stock and then click **Add**. ■

TIP

Conversions The Calculator app also performs conversions from one measure to another. ■

FINDING NEW APPS IN THE WINDOWS STORE

When you're in need of a new app to perform a particular task, the first place to look is in the Microsoft Windows Store. This is an online store for Universal-style apps, both free and paid. You shop the Windows Store by clicking the **Windows Store** item on the Start menu or the taskbar.

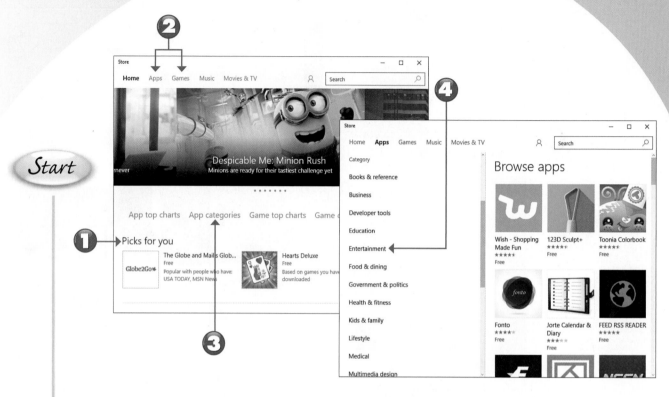

① The Windows Store launches with featured apps at the top of the home page. Scroll down to view Picks for You, Top Free Apps, Best-Rated Apps, New and Rising Apps, Top Free Games, Top Paid Games, Best-Rated Games, New and Rising Games, and Collections.

② To view just applications, click **Apps** at the top of the window. To view just games, click **Games**.

③ Click **App Categories** to view apps by category.

④ Click a category to view all apps in that category.

NOTE

App Store Microsoft's Windows Store is similar in concept to Apple's App Store for iPhones and iPads, as well as the Google Play store for Android devices. ■

TIP

Updating Apps Universal apps you download from the Windows Store do not have to be manually updated. When the publisher improves or upgrades an app, it is automatically updated on your PC. ■

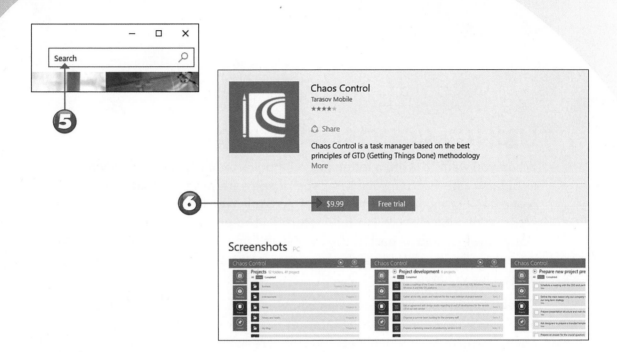

5 To search for apps, enter an app's name into the Search box at the top-right corner of the screen, and then press **Enter**.

6 Click the price button to purchase and install a paid app, or click **Free** to download and install a free app.

End

NOTE

Pricing Whereas a traditional computer software program can cost hundreds of dollars, most apps in the Windows Store cost $10 or less—and many are available free. ■

TIP

Try Before You Buy Most paid apps let you try them before you buy them. Click the **Free Trial** button to install a trial version of that app on your PC. ■

Chapter 8

USING MICROSOFT WORD

When you want to write a letter, fire off a quick memo, create a fancy report, or publish a newsletter, you use a type of software program called a *word processor*. For most computer users, Microsoft Word is the word processing program of choice. Word is a full-featured word processor, and it's included on many new PCs and as part of the Microsoft Office software suite. You can use Word for all your writing needs—from basic letters to fancy newsletters and everything in between.

Several versions of Word are available. Microsoft Word Online is a free web-based version you access using Internet Explorer or another web browser. Then there's the traditional desktop software version of Word, which you can purchase from any consumer electronics store or download from Microsoft or various Internet retailers. Microsoft also makes a universal Word app, that's optimized for touchscreen tablets and phones running Windows. There are even versions of Word optimized for Apple and Android smartphones and tablets.

For many users, Word Online is sufficient, even though it lacks some of the advanced formatting and reference features of the more expensive desktop version. If you want to do sophisticated page layouts, mail merges, and similar functions, you'll need to purchase the desktop software version of Word. Otherwise, use the free online version—it's fine for writing memos, letters, and the like.

COMPARING DESKTOP AND WEB VERSIONS OF WORD

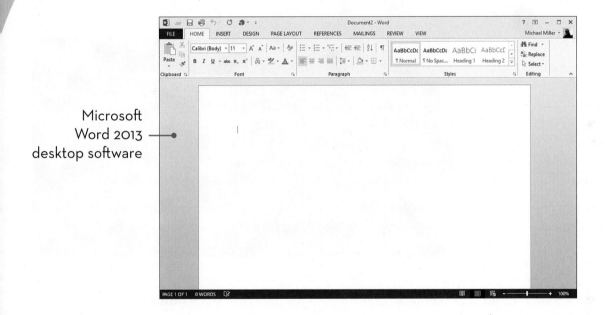

Microsoft
Word 2013
desktop software

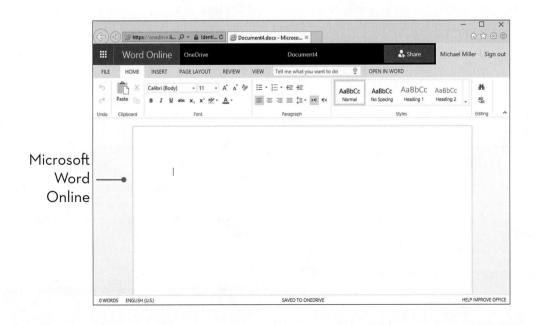

Microsoft
Word
Online

LAUNCHING WORD ONLINE

If you don't want to go to all the trouble of purchasing and installing an expensive piece of software, you can use the Microsoft Word Online free from your web browser.

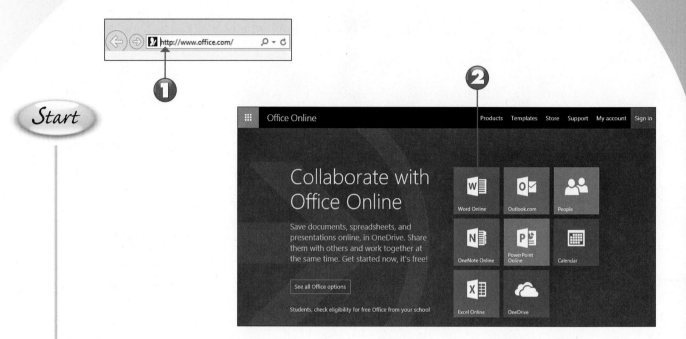

1 From within Internet Explorer or another web browser, enter **www.office.com** into the Address box and press **Enter**.

2 From the Office Online web page, click the **Word Online** tile.

Continued

NOTE

OneDrive You can also open and edit existing Word documents from Microsoft's OneDrive online storage service, located at onedrive.live.com. All the documents you create with Office Online are stored online with OneDrive. ■

NOTE

Microsoft Office Online Microsoft Office Online is a suite of applications that include Word (word processing), Excel (spreadsheets), PowerPoint (presentations), and OneNote (notes and planning). Learn more about Office Online—and the desktop version of Office—online at products.office.com. ■

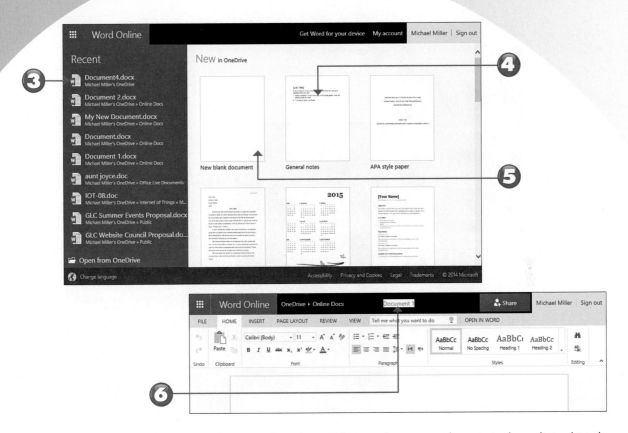

3 You now see the home page for Word Online. Files you've recently created are listed in the Recent pane on the left. To open an existing document, double-click it.

4 To open a new Word document, click one of the templates shown in the main part of the window *or...*

5 Click **New Blank Document** to open a blank document without a template.

6 The document opens with the filename Document 1 or something similar. To change the name of this document, highlight the existing filename at the top of the workspace and enter a new name.

End

NOTE

Templates A template is a document with preformatted styles and often placeholder text. ∎

NOTE

Word for Windows 10 Microsoft offers a universal app version of Word (and of all Office apps), from the Windows Store. This version of Word really isn't designed for Windows 10 computers; it's optimized for use with Windows tablets, smartphones, and other touchscreen devices. You can, however, run it on your Windows 10 PC. ∎

LAUNCHING THE WORD DESKTOP APP

When you need to create more sophisticated documents, use the full-featured desktop version of Microsoft Word. It works similarly to the web version, but with more formatting options.

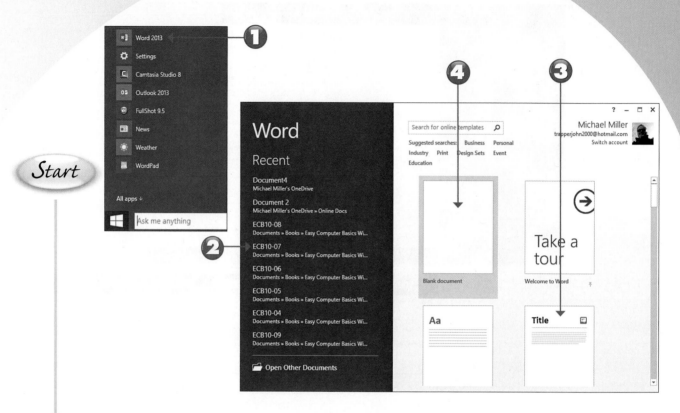

Start

End

1. Open the Windows Start menu and click the **Word 2013** tile or item.

2. Word 2013 launches. Files you've recently created are listed in the Recent pane on the left. To open an existing document, double-click it.

3. To open a new Word document, click one of the templates shown in the main part of the window or...

4. Click **New Blank Document** to open a blank document without a template.

 NOTE

Word 2013 The latest version of Microsoft Word is Word 2013. Older versions look somewhat different and operate slightly differently from what is described in this chapter. ■

 NOTE

Office 365 Home Premium You usually purchase Microsoft Word as part of the Microsoft Office suite of programs. Microsoft offers several editions of Office for purchase, but most home users will find the Home Premium edition the best fit, because it includes the Word, Excel, PowerPoint, Outlook, Publisher, Access, and OneNote apps. You can also get Word (and the rest of Office) on an annual subscription basis; you'll pay $99.99/year to install what Microsoft calls Office 365 on up to five PCs. ■

NAVIGATING WORD ONLINE

Word Online, like the desktop version of Word, uses a ribbon-based interface with different ribbons for different types of operations. Each ribbon contains buttons and controls for specific operations. For example, the Home ribbon contains controls for formatting fonts, paragraphs, and the like; the Insert ribbon includes controls for inserting tables, pictures, clip art, and such.

Start

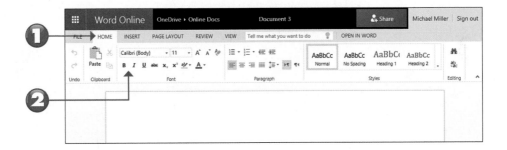

1 Click any tab to display the related ribbon.

2 Click a button or control on the ribbon to perform the given operation.

End

TIP

Context-Sensitive Ribbons Some ribbons appear automatically when you perform a specific task. For example, if you insert a picture and then select that picture, a new Format ribbon tab (not otherwise visible) will appear, with controls for formatting the selected picture. ■

TIP

Different Ribbons The desktop software version of Microsoft Word contains additional ribbons (such as Design, References, and Mailings) not found in Word Online. ■

ENTERING TEXT

You enter text in a Word document at the *insertion point*, which appears onscreen as a blinking cursor. When you start typing on your keyboard, the new text is added at the insertion point.

Start

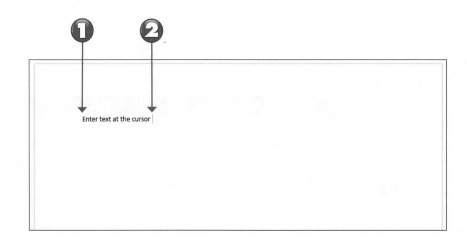

Enter text at the cursor

1 Within your document, click where you want to enter the new text.

2 Type the text.

End

TIP

Move the Insertion Point You move the insertion point with your mouse by clicking a new position in your text. You move the insertion point with your keyboard by using your keyboard's arrow keys. ∎

NOTE

Working with Documents Anything you create with Word—a letter, memo, newsletter, and so on— is called a *document*. A document is nothing more than a computer file that can be copied, moved, deleted, or edited from within Word. ∎

Word lets you cut, copy, and paste text—or graphics—to and from anywhere in your document or between documents. Use your mouse to select the text you want to edit, and then select the appropriate command from the Home ribbon.

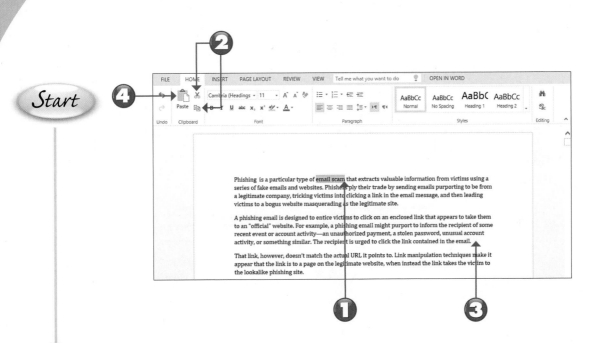

Start

1 Click and drag the cursor to select the text you want to copy or cut.

2 From the Home ribbon, click **Copy** to copy the text or **Cut** to cut the text.

3 Within the document, click where you want to paste the cut or copied text.

4 From the Home ribbon, click **Paste**.

End

TIP

Keyboard Shortcuts You also can select text using your keyboard; use the Shift key—in combination with other keys—to highlight blocks of text. For example, Shift + left arrow selects one character to the left. ■

NOTE

Cut Versus Copy Cutting text removes the text from the original location, at which point you can paste it into a new location. When you copy and paste text, the text stays in the original location and a copy of it is placed into a new location—essentially duplicating the text. ■

FORMATTING TEXT

After your text is entered and edited, you can use Word's numerous formatting options to add some pizzazz to your document.

Font   Font size   Font color

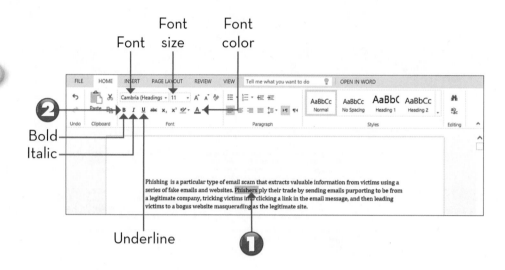

Bold

Italic

Underline

1 Click and drag the cursor over the text you want to edit.

2 Click the desired button in the Font section of the Home ribbon: **Font**, **Font Size**, **Bold**, **Italic**, **Underline**, or **Font Color**.

FORMATTING PARAGRAPHS

When you're creating a more complex document, you need to format more than just a few words here and there. To format complete paragraphs, use Word's Paragraph formatting options on the Home ribbon.

Bullets — Numbering — Decrease Indent — Increase Indent

Start

Align Text Left

Center

Align Text Right

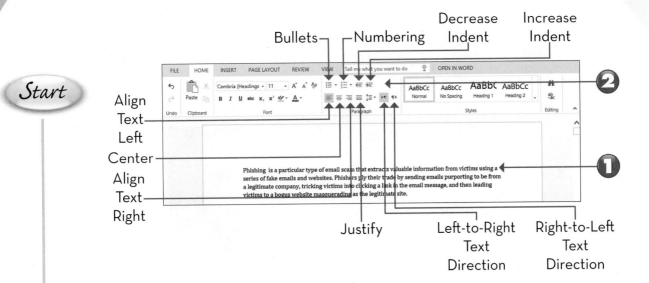

Justify — Left-to-Right Text Direction — Right-to-Left Text Direction

1 Click anywhere within the paragraph you want to format.

2 Click the desired button in the Paragraph section of the Home ribbon—including **Bullets**, **Numbering**, **Decrease Indent**, **Increase Indent**, **Line Spacing**, or any of the **Align Text** options.

End

TIP

Spell Checking If you misspell a word, it appears onscreen with a squiggly red underline. Right-click the misspelled word and select the correct spelling from the list. ■

SAVING YOUR WORK

If you're working on a file in the Word desktop app, you need to save your edits periodically. This is an easy process.

Start

1 Select the **File** ribbon to display the Info panel.

2 Click **Save**.

End

TIP

Saving Your Work Online If you're using Word Online, you don't have to manually save your work. The online app automatically saves any changes you make—so you can ignore the instructions on this page! ▪

PRINTING A DOCUMENT

When you've finished editing your document, you can instruct Word to send a copy to your printer.

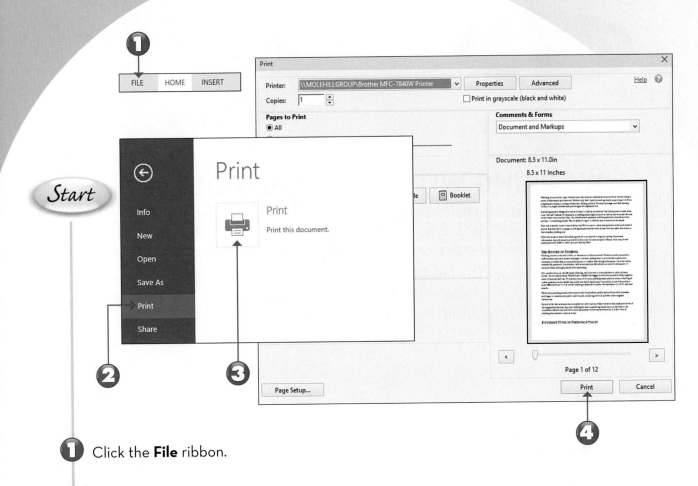

1 Click the **File** ribbon.

2 Click **Print** to display the Print pane.

3 Click the **Print** button to display the **Print** panel.

4 Configure any necessary options, and then click the **Print** button to print the document.

Chapter 9

WORKING WITH FILES AND FOLDERS

All the data for documents and programs on your computer is stored in electronic files. These files are then arranged into a series of folders and subfolders—just as you'd arrange paper files in a series of file folders in a filing cabinet.

In Windows 10, you use File Explorer to view and manage your folders and files. You can navigate to folders and files on your computer and connected devices, or to those on other computers on your network.

FILE EXPLORER

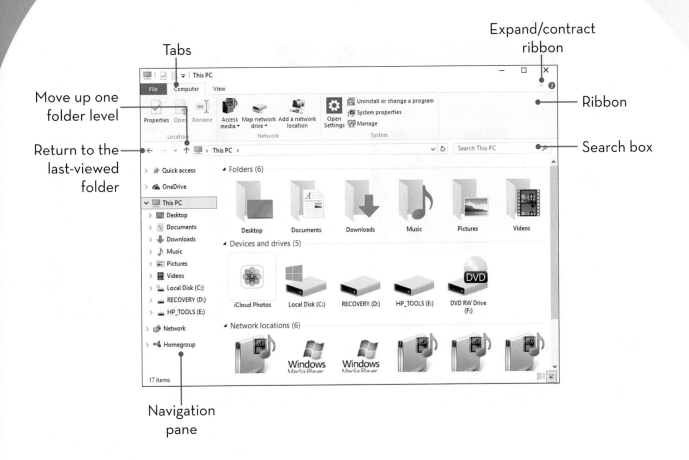

Expand/contract ribbon

Tabs

Move up one folder level

Return to the last-viewed folder

Ribbon

Search box

Navigation pane

LAUNCHING FILE EXPLORER

There are three ways to open File Explorer in Windows 10.

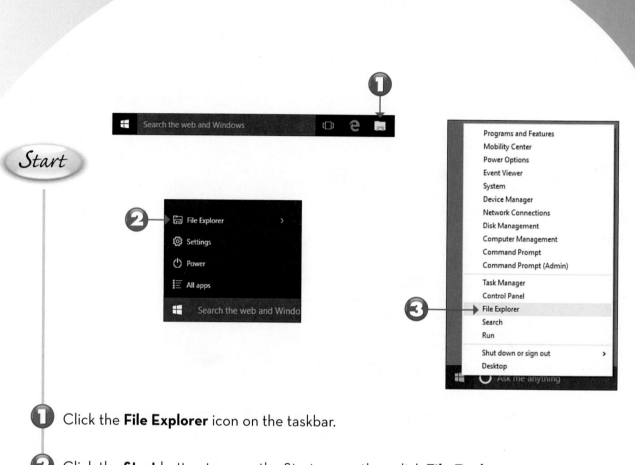

Start

End

1 Click the **File Explorer** icon on the taskbar.

2 Click the **Start** button to open the Start menu, then click **File Explorer**.

3 Right-click the Start menu to display the Quick Access menu, and then click **File Explorer**.

NOTE

Windows Explorer In some previous versions of Windows, File Explorer was known as Windows Explorer—or, more colloquially, as either the My Computer or My Documents folder. ■

EXPLORING THE RIBBON

Most file-related operations are located on the ribbon at the top of the File Explorer window. The File Explorer ribbon consists of three tabs and the File drop-down menu.

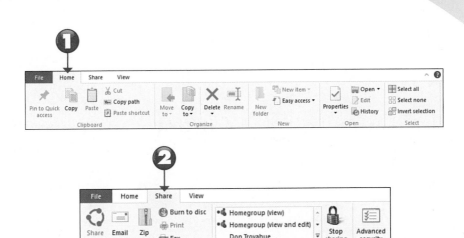

 Start

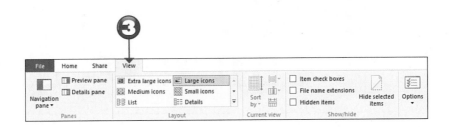

1 Select the **Home** tab to access the most common file-related commands—Cut, Copy, Paste, Move, New Folder, and so forth.

2 Select the **Share** tab to share or email a folder or file, burn files to a removable disc, print a file, or "zip" files into a compressed folder.

3 Select the **View** tab to change the way files and folders appear in File Explorer, as well as sort and group items.

End

NOTE

File Menu Click the File menu to open a new window, the DOS command prompt (used on older operating systems), or the Windows PowerShell (a more advanced version of the command prompt). You can also use the File menu to delete recent history, open the Help system, or close the File Explorer window. ■

NAVIGATING WITH THE NAVIGATION PANE

The Navigation pane on the left side of the File Explorer window displays both favorite links and hierarchical folder trees for your computer, network, and OneDrive online storage. Click the arrow icon next to any folder to display all the subfolders it contains. Click a folder to display its contents in the main File Explorer window.

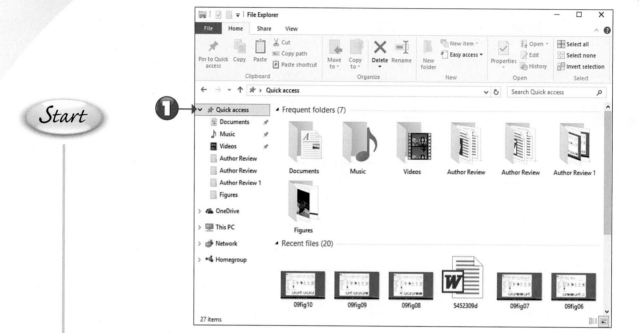

Start

1 Select **Quick Access** to view frequently accessed folders and recent files.

Continued

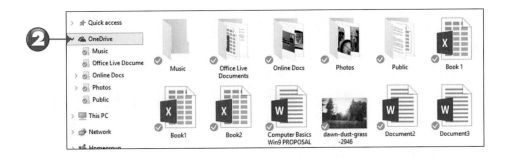

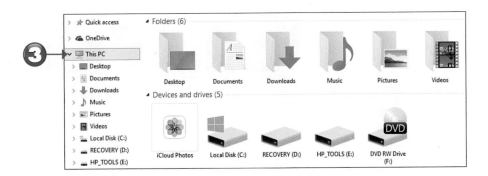

2 Select **OneDrive** to view files and folders stored online on Microsoft's OneDrive cloud storage service.

3 Select **This PC** to view your default folders (Desktop, Documents, Downloads, Music, Pictures, and Videos), as well as all drives and devices connected to your PC.

Continued

 NOTE

Major Folders The Documents, Music, Pictures, and other folders displayed when you click **This PC** are not the only folders on your PC, but they are the ones most likely to contain data files. Other folders, which can be displayed if you click the **Local Disk** icon, are more likely to contain programs and apps rather than documents and data. ∎

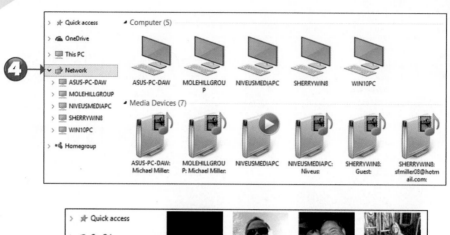

4 Select **Network** to view all computers connected to your network.

5 Select **Homegroup** to view all users and computers connected to your network homegroup.

End

WORKING WITH FOLDERS

You can navigate through the folders and subfolders in File Explorer in several ways.

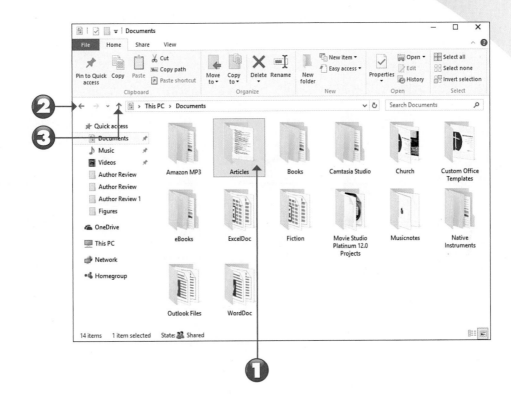

Start

1 A given folder might contain multiple subfolders. Double-click any item to view its contents.

2 To move back to the disk or folder previously selected, click the **Back** button on the toolbar.

3 To move up the hierarchy of folders and subfolders to the next highest item, click the **up arrow** button on the toolbar.

End

TIP

Breadcrumbs The list of folders and subfolders in File Explorer's Address box presents a "breadcrumb" approach to navigation. You can view additional items by clicking the **separator arrow** next to the folder icon in the Address box; this displays a pull-down menu of the contents of the item to the left of the arrow. ■

CHANGING THE WAY FILES ARE DISPLAYED

You can choose to view the contents of a folder in various ways. The icon views are nice in that they show a small thumbnail preview of any selected file.

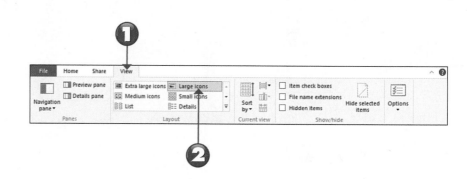

1 Click the **View** tab on the ribbon.

2 Go to the Layout section and click one of the following view options: **Content**, **Tiles**, **Details**, **List**, **Small Icons**, **Medium Icons**, **Large Icons**, or **Extra Large Icons**.

End

TIP

Which View Is Best? Any of the larger icon views is best for working with graphics files. Details view is best if you're looking for files by date or size. ■

SORTING FILES AND FOLDERS

When viewing files in File Explorer, you can sort your files and folders in a number of ways. To view your files in alphabetic order, choose to sort by **Name**. To see all similar files grouped together, choose to sort by **Type**. To sort your files by the date and time they were last edited, select **Date Modified**.

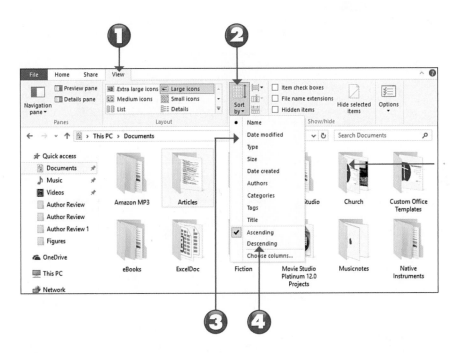

① Click the **View** tab on the ribbon.

② Click the **Sort By** button.

③ Choose to sort by **Name**, **Date Modified**, **Type**, **Size**, **Date Created**, **Authors**, **Categories**, **Tags**, or **Title**.

④ By default, Windows sorts items in ascending order. To change the sort order, click **Descending**.

End

TIP

Different Sorting Options Different types of files have different sorting options. For example, if you're viewing music files, you can sort by **Album**, **Artists**, **Bit Rate**, **Composers**, **Genre**, and the like. ∎

CREATING A NEW FOLDER

The more files you create, the harder it is to organize and find things on your hard disk. When the number of files you have becomes unmanageable, you need to create more folders—and subfolders—to better categorize your files.

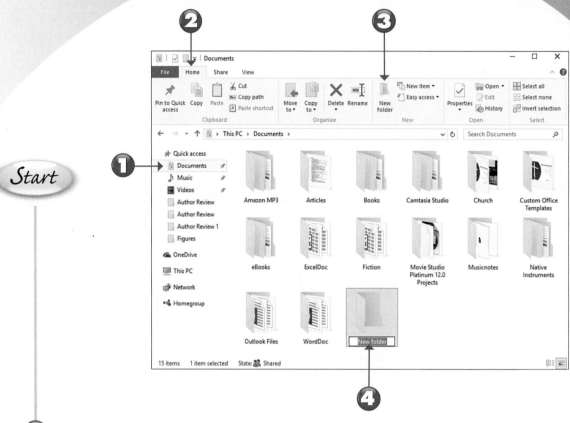

Start

1. Navigate to the drive or folder where you want to place the new folder.

2. Click the **Home** tab on the ribbon.

3. Click the **New Folder** button.

4. A new, empty folder now appears with the filename New Folder highlighted. Type a name for your folder and press **Enter**.

End

CAUTION

Illegal Characters Folder names and filenames can include up to 255 characters—including many special characters. You *can't*, however, use the following "illegal" characters: \ / : * ? " < > |. ▪

RENAMING A FILE OR FOLDER

When you create a new file or folder, it helps to give it a name that describes its contents. Sometimes, however, you might need to change a file's name. Fortunately, Windows makes renaming an item relatively easy.

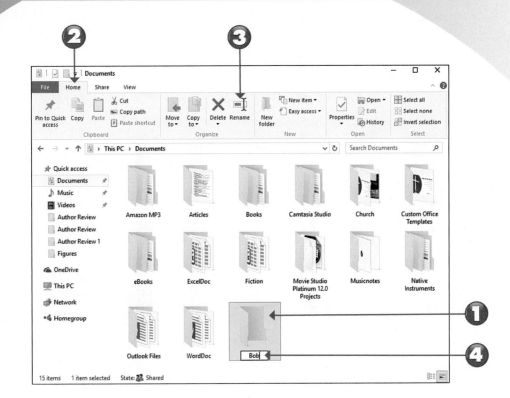

Start

① Click the file or folder you want to rename.

② Click the **Home** tab on the ribbon.

③ Click the **Rename** button; this highlights the filename.

④ Type a new name for your folder (which overwrites the current name), and press **Enter**.

End

CAUTION

Don't Change the Extension The one part of the filename you should never change is the extension—the part that comes after the final "dot" if you choose to show file extensions. Try to change the extension, and Windows will warn you that you're doing something wrong. ■

TIP

Keyboard Shortcut You can also rename a file by selecting the file and pressing **F2** on your computer keyboard. This highlights the filename and readies it for editing. ■

COPYING A FILE OR FOLDER

There are many ways to copy a file in Windows. The easiest method is to use the **Copy To** button on the Home ribbon.

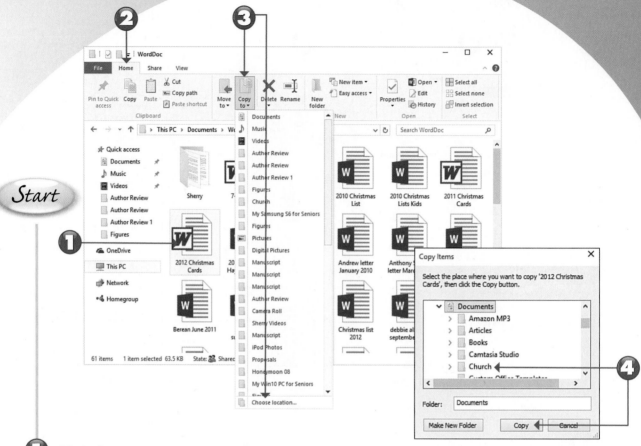

Start

1 Click the item you want to copy.

2 Click the **Home** tab on the ribbon.

3 Click the **Copy To** button and select one of the suggested locations or select **Choose Location** to copy the file elsewhere.

4 When the Copy Items dialog box appears, navigate to the new location for the item and then click the **Copy** button.

End

MOVING A FILE OR FOLDER

Moving a file or folder is different from copying it. Moving cuts the item from its previous location and pastes it into a new location. Copying leaves the original item where it was *and* creates a copy of the item elsewhere.

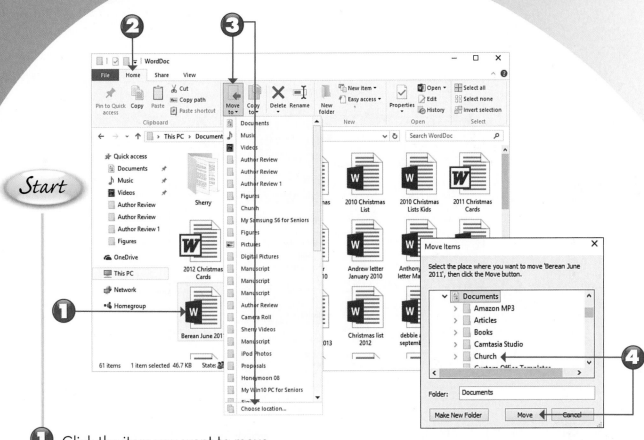

1. Click the item you want to move.

2. Click the **Home** tab on the ribbon.

3. Click the **Move To** button and select one of the suggested locations or click **Choose Location** to move the file to another location.

4. When the Move Items dialog box appears, navigate to the new location for the item, and then click the **Move** button.

End

SEARCHING FOR A FILE

As organized as you might be, you might not always be able to find the specific files you want. Fortunately, Windows offers an easy way to locate difficult-to-find files, via the Instant Search function. Instant Search lets you find files by extension, filename, or keywords within the file.

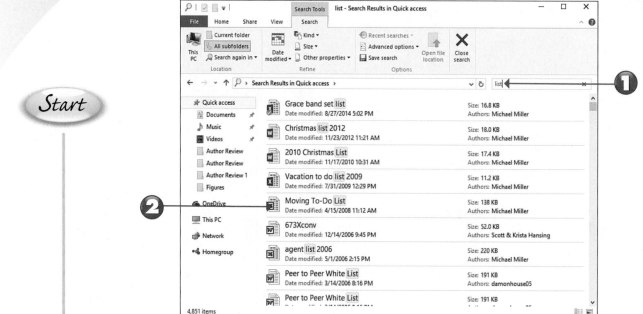

Start

1 From within File Explorer, enter one or more keywords into the Search box and press **Enter**.

2 Windows now displays a list of files that match your search criteria. Double-click any item to open that file.

End

TIP

Search Index Instant Search indexes all the files stored on your hard disk (including email messages) by type, title, and contents. ■

DELETING A FILE OR FOLDER

Keeping too many files eats up too much hard disk space—which is a bad thing. Because you don't want to waste disk space, you should periodically delete those files (and folders) you no longer need. When you delete a file, you send it to the Windows Recycle Bin, which is kind of a trash can for deleted files.

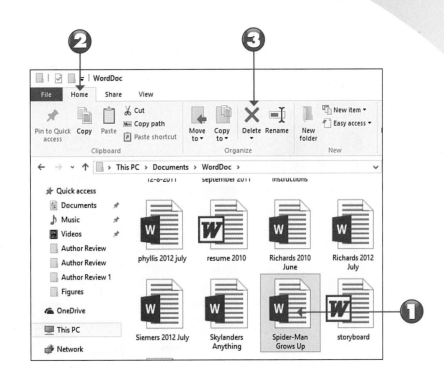

Start

1 Click the file you want to delete.

2 Click the **Home** tab on the ribbon bar.

3 Click the **Delete** button.

End

RESTORING DELETED FILES

Have you ever accidentally deleted the wrong file? If so, you're in luck. Windows stores the files you delete in the Recycle Bin, which is actually a special folder on your hard disk. For a short time (in most instances, several days), you can "undelete" files from the Recycle Bin back to their original locations.

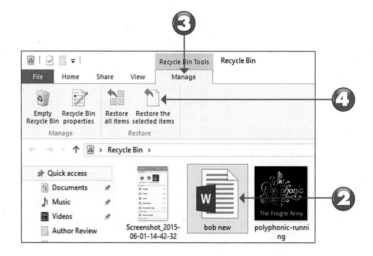

Start

① From the Windows desktop, double-click the **Recycle Bin** icon to open the Recycle Bin folder.

② Click the file you want to restore.

③ Click the **Manage** tab on the ribbon bar.

④ Click the **Restore the Selected Items** button.

End

EMPTYING THE RECYCLE BIN

By default, the deleted files in the Recycle Bin can occupy 4GB plus 5% of your hard disk space. When you've deleted enough files to exceed this limit, the oldest files in the Recycle Bin are automatically and permanently deleted from your hard disk. You can also manually empty the Recycle Bin and thus free up some hard disk space.

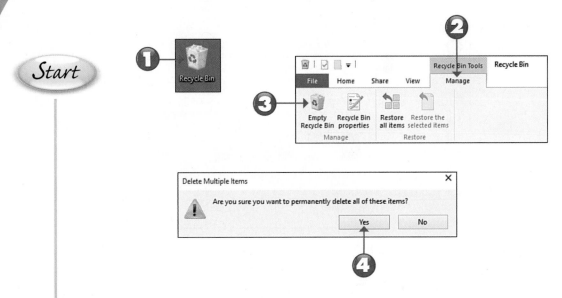

1 From the Windows desktop, double-click the **Recycle Bin** icon to open the Recycle Bin folder.

2 Click the **Manage** tab on the ribbon bar.

3 Click the **Empty the Recycle Bin** button.

4 When the Delete Multiple Items dialog box appears, click **Yes** to completely erase the files.

TIP

Fast Empty You can also empty the Recycle Bin by right-clicking its icon on the Windows desktop and selecting **Empty Recycle Bin** from the pop-up menu. ■

COMPRESSING A FILE

Really big files can be difficult to copy or share. Fortunately, Windows lets you create *compressed* folders, which take big files and compress them in size (called a "zipped" file). After the file has been transferred, you can then uncompress the file to its original state.

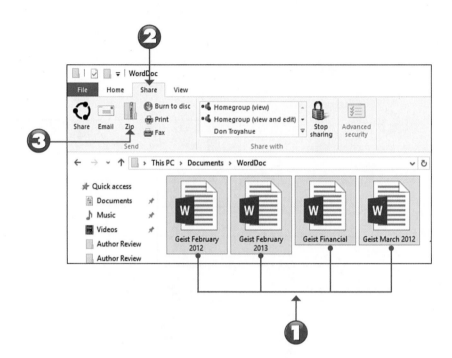

1. Click the files you want to compress. (To select more than one file, hold down the **Ctrl** key while clicking.)

2. Click the **Share** tab on the ribbon bar.

3. Click the **Zip** button. Windows now creates a new folder that contains compressed versions of the files you selected.

End

NOTE

Zip Files The compressed folder is actually a file with a .zip extension, so it can be used with other compression/decompression programs. ■

EXTRACTING FILES FROM A COMPRESSED FOLDER

The process of decompressing a file is actually an *extraction* process. That's because you extract the original files from the compressed folder to the desired location on your computer's hard drive.

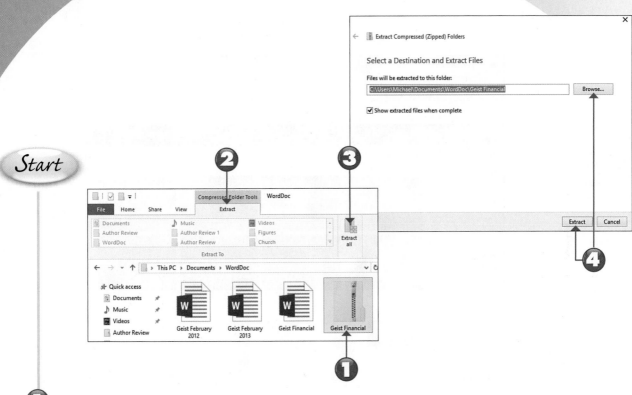

 Select the compressed folder.

Click the **Extract** tab on the ribbon bar.

Click the **Extract All** button.

 When the Extract Compressed (Zipped) Folders dialog box appears, click the **Browse** button to select a location for the extracted files, and then click the **Extract** button.

TIP
Extracted Folder By default, compressed files are extracted to a new folder with the same name. You can change this, however, to extract to any folder you like. ■

TIP
Zipper Icon Compressed folders are distinguished by the little zipper on the folder icon. ■

WORKING WITH FILES ON ONEDRIVE

Microsoft offers online storage for all your documents and data, via its OneDrive service. When you store your files on OneDrive, you can access them via any computer or mobile device connected to the Internet. You manage all your online files from the OneDrive website, using your web browser.

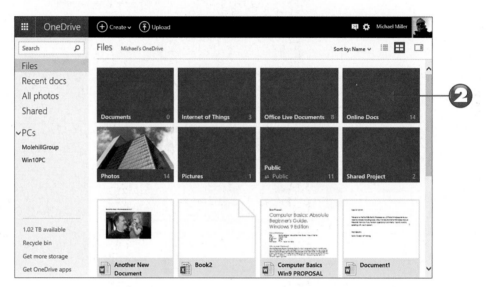

① From within your web browser, enter **onedrive.live.com** into the Address box and press **Enter**.

② Your OneDrive files are stored in folders. Click a folder to view its contents.

Continued

 TIP

Syncing Folders You can synchronize folders between your computer and the OneDrive service. From within File Explorer, move a given folder into the OneDrive folder. That folder now appears on the OneDrive website and on your PC. Make changes in one location and those changes are automatically applied to the other. Click OneDrive in File Explorer's Navigation pane to view your synced folders. ■

 NOTE

Cloud Storage Online file storage, such as that offered by OneDrive, Apple's iCloud, and Google Drive, is called *cloud storage*. The main advantage of cloud storage is that files can be accessed from any computer (work, home, or other) at any location. You're not limited to using a given file on a single computer only. ■

3 Click a file to view it or, in the case of an Office document, open it in its host application.

4 To copy, move, or rename a file, right-click the file to display the options menu, and then select the action you want to perform.

End

TIP

Download a File To download a file from OneDrive to your local hard disk, right-click the file to display the options menu, and then click **Download**. ◾

NOTE

Storage Plans Microsoft gives you 15GB of storage in your free OneDrive account, which is more than enough to store most users' documents, digital photos, and the like. If you need more storage, you can purchase 100GB of storage for $1.99/month, 200GB storage for $3.99/month, or 1TB storage for $6.99/month. ◾

USING THE INTERNET

It used to be that most people bought personal computers primarily to do work with productivity software—word processing, spreadsheets, databases, that sort of thing. But today, many people also buy PCs to access the Internet—to send and receive email, surf the Web, and keep in touch with Facebook and other social media.

If you're using your notebook or tablet PC on the road, accessing the Internet is as easy as finding a public Wi-Fi hotspot, like the kind offered at Starbucks, McDonald's, and other businesses. Of course, if you're at home, you access the Internet from your home wireless network. You connect your computer to the hotspot or wireless network, which then connects you to the Internet. Easy.

However you go online, you use a *web browser*, such as Microsoft Edge or Google Chrome, to browse that part of the Internet called the World Wide Web. (Or just Web, for short.) Information on the Web is presented in *web pages*, each of which contains text, graphics, and links to other web pages. A web page resides at a *website*, which is nothing more than a collection of web pages. The main page of a website is called the *home page*, which serves as an opening screen that provides a brief overview and a sort of menu of everything you can find at that site.

MICROSOFT EDGE

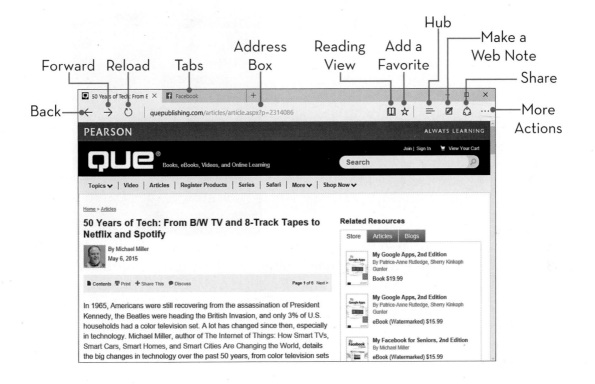

CONNECTING TO AN INTERNET WI-FI HOTSPOT

If you have a notebook PC, you have the option to connect to the Internet when you're out and about. Many coffeehouses, libraries, hotels, and public spaces offer wireless Wi-Fi Internet service, either free or for an hourly or daily fee.

Start

① On the taskbar, click the **Connections** button to display the Connections pane.

② You see a list of available wireless networks. Click the network to which you want to connect.

Continued

NOTE

Wireless Hotspots A *hotspot* is a public place that offers wireless access to the Internet using Wi-Fi technology. Some hotspots are free for all to access; others require some sort of payment. ■

NOTE

Finding the Wi-Fi Signal When you're near a Wi-Fi hotspot, your PC should automatically pick up the Wi-Fi signal. Just make sure that the Wi-Fi adapter in your computer is turned on (some notebooks have a switch for this, either on the front or on the side of the unit), and then follow these directions to find and connect to the nearest hotspot. ■

3 This expands the panel; click **Connect** to connect to the selected hotspot.

4 If the hotspot has free public access, you can now open your web browser and surf normally. If the hotspot requires a password, payment, or some other logon procedure, Windows should automatically open your web browser and display the hotspot's logon page. Enter the appropriate information to begin surfing.

End

TIP

Airplane Mode If you're using your notebook or tablet on an airplane and don't want to use the plane's wireless Internet service (if available), you can switch to Airplane mode so that you can use your computer while in the air. To switch into Airplane mode, click the **Connections** button on the taskbar to open the Connections pane, and then click the **Airplane Mode** tile. You can switch off Airplane mode when your plane lands. ■

WEB BROWSING WITH MICROSOFT EDGE

You can use any web browser to visit websites online. Windows 10 comes with a new web browser called Microsoft Edge. You can also use browsers from other companies, including Google Chrome and Mozilla Firefox. They all work in similar ways, but this chapter's examples use Microsoft Edge.

1 Open Microsoft Edge and enter a web page address into the **Address** box. (You may need to click near the top of the browser to display the Address box.)

2 As you type, Edge displays a list of suggested pages. Click one of these pages or finish entering the web page address and press **Enter**.

Continued

TIP

Downloading Other Web Browsers Most other web browsers are free and can be easily downloaded over the Internet (using Microsoft Edge or another web browser). To download Google Chrome, go to www.google.com/chrome/. To download Mozilla Firefox, go to www.mozilla.org/firefox/. ■

TIP

Start Page Microsoft Edge lets you set a *start page* that automatically opens whenever you launch the browser. (Edge's default Start page displays a list of your Top Sites and a Bing search box.) To set a new start page, click the **More Actions** (three-dot) button and click **Settings**. When the Settings pane appears, go to the Open With section and select **A Specific Page or Pages**. Click the page list, select **Custom**, and then enter the URL of the page you want into the **Enter a Web Address** box. ■

3 To return to the previous web page, click the **Back** (left arrow) button beside the Address box.

4 To reload or refresh the current page, click the **Refresh** button.

5 Pages on the Web are linked via clickable *hyperlinks*. Click a link to display the linked-to page.

End

NOTE

Internet Explorer Microsoft Edge replaces the Internet Explorer web browser, which was the default browser included with previous versions of Windows. Edge is a faster and more modern browser, compatible with more of today's state-of-the-art websites. ■

TIP

Revisit History To view a list of pages you've recently visited, click the **Hub** (three-line) button and select the **History** (timer) tab. Click the page you want to revisit. ■

SAVING FAVORITE PAGES

All web browsers let you save or bookmark your favorite web pages. In Microsoft Edge, you do this by adding pages to the Favorites list.

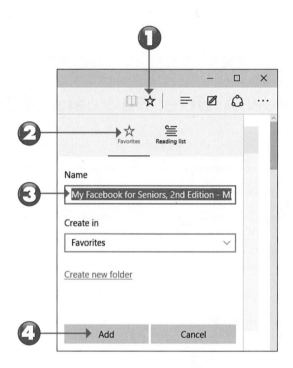

Start

End

Navigate to the web page you want to add to your Favorites list, and then click the **Favorites** (star) icon in the Address box.

Click to select the **Favorites** tab.

Confirm or enter a name for this page.

Click the **Add** button.

TIP

Favorites Folders You can organize your favorite pages into separate folders in the Favorites list. After you click the Favorites icon, select a folder from the Create In list or click **Create New Folder** to create a new folder. ■

TIP

Reading View Some web pages are overly cluttered with advertisements and other distracting elements. To display a web article with just text and accompanying pictures, click the **Reading View** button on the Edge toolbar. This removes the unnecessary elements and makes reading easier. ■

RETURNING TO A FAVORITE PAGE

To return to a page you've saved as a favorite, open the Favorites list and make a selection.

Start

 Click the **Hub** button to display the Hub panel.

 Click the **Favorite** (star) tab to display your Favorites list.

Click the page you want to revisit.

End

TIP

Favorites Bar For even faster access to your favorite pages, display the Favorites bar at the top of the browser window, under the Address bar. Click the **More Actions** button and select **Settings** to display the Settings page, and then set the **Show the Favorites Bar** control to the On position. ◼

TIP

Web Notes If you're doing research on the Web, you might want to mark up a given web page with notes, highlights, and such. You can do this with Edge's Web Notes tool. Click the **Make a Web Note** button to display the Web Notes toolbar; use the appropriate markup tool to make your notes. ◼

OPENING MULTIPLE PAGES IN TABS

Most web browsers, including Microsoft Edge, let you display multiple web pages as separate tabs, and thus easily switch between web pages. This is useful when you want to reference different pages or want to run web-based applications in the background.

Start

1 To open a new tab, click the **+** next to the last open tab.

2 To switch tabs, click the tab you want to view.

3 Click the X on the current tab to close it.

End

TIP

Anonymous Browsing If you want to browse anonymously, without any traces of your history recorded, activate Edge's InPrivate Browsing mode in a new browser window. Click the **More Actions** button and then select **New InPrivate Window**. ■

SEARCHING THE WEB WITH GOOGLE

You can find just about anything you want online by using a web *search engine*. The most popular search engine today is Google (www.google.com), which indexes billions of individual web pages. Google is very easy to use and returns extremely accurate results.

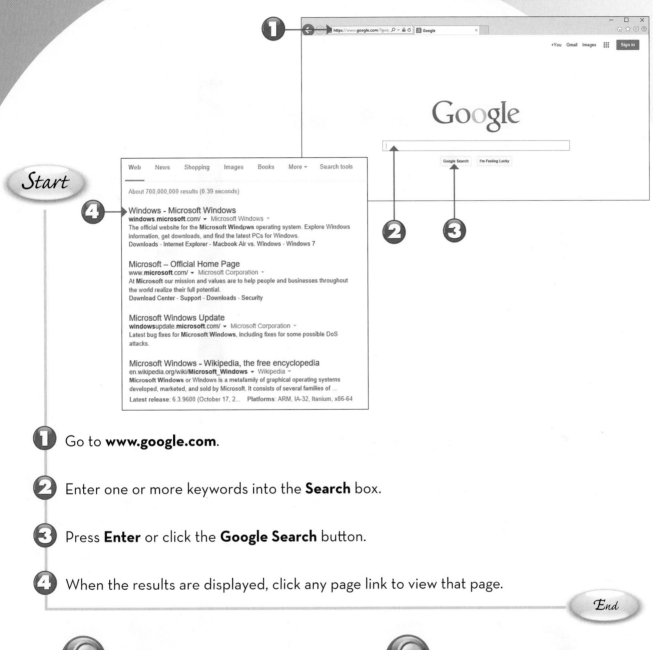

Web   News   Shopping   Images   Books   More ▾   Search tools

About 700,000,000 results (0.39 seconds)

Windows - Microsoft Windows
windows.**microsoft**.com/ ▾ Microsoft Windows ▾
The official website for the **Microsoft Windows** operating system. Explore Windows information, get downloads, and find the latest PCs for Windows.
Downloads - Internet Explorer - Macbook Air vs. Windows - Windows 7

Microsoft – Official Home Page
www.**microsoft**.com/ ▾ Microsoft Corporation ▾
At **Microsoft** our mission and values are to help people and businesses throughout the world realize their full potential.
Download Center - Support - Downloads - Security

Microsoft Windows Update
windowsupdate.**microsoft**.com/ ▾ Microsoft Corporation ▾
Latest bug fixes for **Microsoft Windows**, including fixes for some possible DoS attacks.

Microsoft Windows - Wikipedia, the free encyclopedia
en.wikipedia.org/wiki/**Microsoft_Windows** ▾ Wikipedia ▾
Microsoft Windows or Windows is a metafamily of graphical operating systems developed, marketed, and sold by Microsoft. It consists of several families of ...
Latest release: 6.3.9600 (October 17, 2...   **Platforms**: ARM, IA-32, Itanium, x86-64

Start

1 Go to **www.google.com**.

2 Enter one or more keywords into the **Search** box.

3 Press **Enter** or click the **Google Search** button.

4 When the results are displayed, click any page link to view that page.

End

TIP

Fine-Tune Search Results You can fine-tune your results by using the search tools located at the top of the search results page. Click **Search Tools** and select to filter by time, location, and other criteria. ■

TIP

Microsoft Bing Microsoft offers its own competing search engine, dubbed Bing. You can search Bing at www.bing.com. Bing also powers the search within the Cortana virtual assistant, which is discussed in the next section. ■

SMART SEARCHING WITH CORTANA

Windows 10 includes a virtual personal assistant, called *Cortana*, that can help you search for files on your computer and information on the Web. You can query Cortana by typing into the search box on the taskbar, or with voice commands by speaking into your computer's microphone.

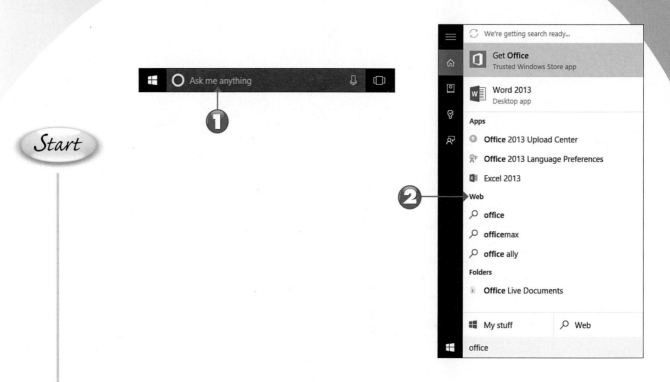

Start

1. Click within the **Ask Me Anything** search box in the taskbar and begin typing your query.

2. As you type, Cortana displays items that match your query in the results pane. This might include apps and files on your computer, as well as results from the Web. Click an item to display or open it.

Continued

TIP
Voice Commands If your computer has a built-in microphone, or if you have a microphone connected to your computer, you can control Cortana with voice commands. Click the microphone within the Cortana search box or speak, "Hey, Cortana," into your computer's microphone, followed by whatever it is you're asking. ■

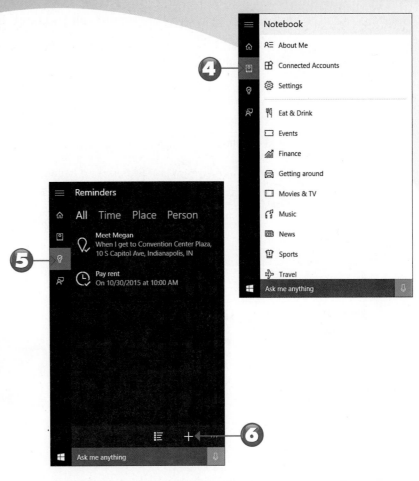

 3 To display news items, stock prices, weather conditions, and other items personal to you, click within the search box to display the Cortana pane.

4 To personalize which content Cortana displays, click the **Notebook** icon and then click the content you want to customize.

5 To see reminders, click the **Reminders** tab.

6 Click the **+** to set a new reminder.

End

TIP
Virtual Assistant The more you use Cortana, the more it will learn about you and display information you find useful. You can use Cortana to track weather conditions, traffic conditions, stock prices, airline flights, and more. ■

TIP
Cortana in Edge You can also use Cortana to display more information about a given topic from within the Edge browser. Highlight a word or phrase on a web page, right-click, and select **Ask Cortana**. You now see a Cortana pane within Edge displaying additional information. ■

SHOPPING ONLINE

The Internet is a great place to buy things, from books to clothing to household items to cars. Online shopping is safe and convenient—all you need is your computer and a credit card.

Start

① Find an online store that sells the item you're shopping for.

② Search or browse for the product you like.

Continued

TIP

Traditional Retailers Online Most brick-and-mortar retailers have equivalent online stores. For example, you can shop at Target online at www.target.com, or Macy's online at www.macys.com. Most catalog merchants also have their own websites where you can order online. ▪

TIP

Online-Only Retailers Many big online-only retailers sell a variety of merchandise. The most popular of these include Amazon.com (www.amazon.com) and Overstock.com (www.overstock.com). ▪

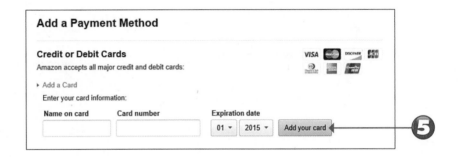

3 Examine the product by viewing the photos and information on the product listing page.

4 Order the product by clicking a **Buy It Now** or **Add to Cart** button on the product listing page. This puts the item in your online shopping cart.

5 Check out by entering your shipping and payment (credit card) information.

End

TIP

In-Stock Items The better online retailers tell you either on the product description page or during the checkout process whether an item is in stock. Look for this information to help you decide how to group your items for shipment. ■

TIP

Shop Safely The safest way to shop online is to pay via credit card, because your credit card company offers various consumer protections. (Smaller merchants might accept credit cards via PayPal or a similar online payment service; this is also acceptable.) Also make sure that the retailer you buy from has an acceptable returns policy, just in case. ■

BUYING ITEMS ON CRAIGSLIST

When you're looking to buy something locally, you can often find great bargains on Craigslist (www.craigslist.org), an online classified advertising site. Browse the ads until you find what you want, and then arrange with the seller to make the purchase.

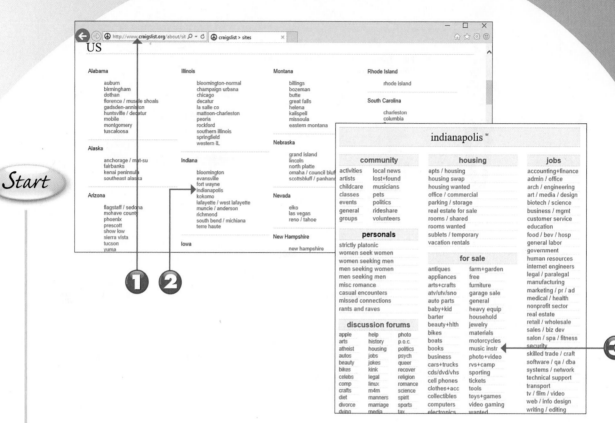

Start

① Go to **www.craigslist.org**.

② Click the name of your city or state.

③ Go to the **For Sale** section and click the category you're looking for.

Continued

NOTE

Classified Ads Listings on Craigslist are just like traditional newspaper classified ads. All transactions are between you and the seller; Craigslist is just the "middleman." ■

TIP

Contacting the Seller When you contact the seller via email, let him know you're interested in the item and would like to see it in person. The seller should reply with a suggested time and place to view and possibly purchase the item. ■

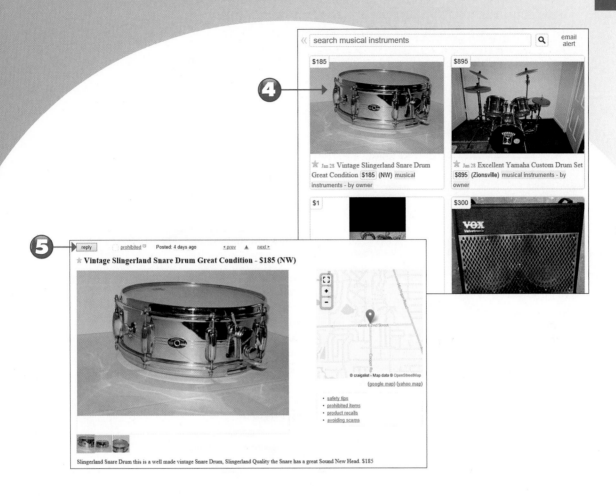

4 Click the link or picture for the item you're interested in.

5 Read the item details, and then click the **Reply** button to email the seller and express your interest.

End

TIP

Pay in Cash When you purchase an item from a Craigslist seller, expect to pick up the item in person and pay in cash. ■

CAUTION

Buyer Beware Just as with traditional classified ads, Craigslist offers no buyer protections. Make sure you inspect the item before purchasing! ■

SELLING ITEMS ON CRAIGSLIST

The Craigslist site is also a great place to sell items you want to get rid of. Just place an ad and wait for potential buyers to contact you!

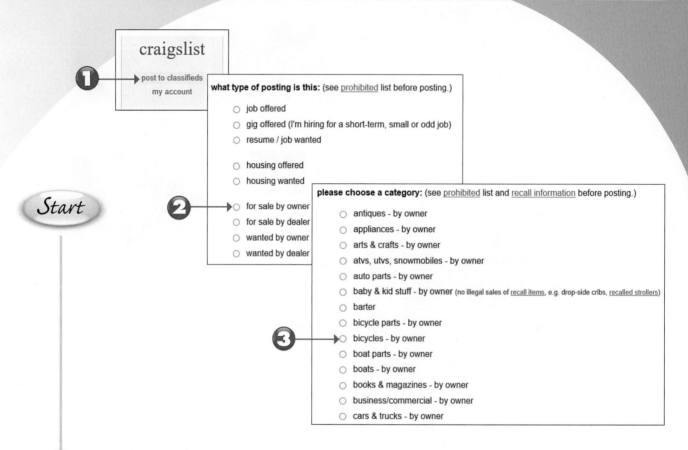

Start

① From the Craigslist site, click the **Post to Classifieds** link.

② Click the type of ad you want to place—typically **For Sale by Owner**.

③ Click the category that best fits what you're selling. (If necessary, click through to an appropriate subcategory.)

Continued

NOTE

Location Depending on where you live, you might be prompted to select a more exact location than just the larger metropolitan area. ■

NOTE

Contact Email For your protection, Craigslist displays an anonymized email address in your item listing. Buyers email this anonymous address and the emails are forwarded to your real email address. ■

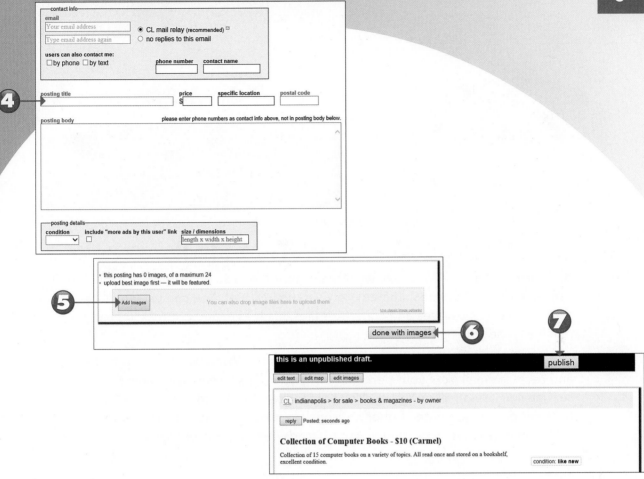

Enter the necessary details about what you're selling, including the listing title, asking price, and description, and then click **Continue**.

You are now prompted to add pictures of your item. (Items sell better if buyers can see what's for sale, although such photos are optional.) Click the **Add Images** button to select digital photos of your item.

Click the **Done with Images** button.

Confirm the listing details, and then click the **Publish** button to finalize the listing.

End

CAUTION

Safety First Make sure someone else is with you before you invite potential buyers into your home to look at the item you have for sale—or arrange to meet buyers at a safe neutral location. ■

TIP

Other Services The Craigslist site isn't just for buying and selling merchandise. You can also use Craigslist to look for or offer services, jobs, and housing. ■

COMMUNICATING WITH EMAIL

An email message is like a regular letter, except that it's composed electronically and delivered almost immediately via the Internet. You can use email to send both text messages and computer files (such as digital photos) to pretty much anyone who's online.

You can use a dedicated email program, such as the Windows Mail app, to send and receive email from your personal computer. Or you can use a web mail service such as Gmail or Yahoo! Mail to manage all your email from any web browser on any computer. Either approach is good and lets you create, send, and read email messages from all your friends, family, and colleagues.

WINDOWS MAIL APP

Reply to
current
message

Delete
current
message

Create new
message

Navigation
pane

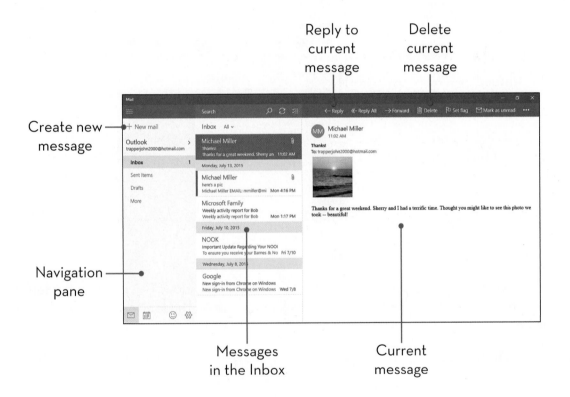

Messages
in the Inbox

Current
message

VIEWING YOUR INBOX AND READING MESSAGES

Windows 10 includes a built-in Mail app for sending and receiving email messages. By default, the Mail app manages email from the Outlook.com or Hotmail account linked to your Microsoft Account. This means you'll see Outlook and Hotmail messages in your Mail Inbox and be able to easily send emails from your Outlook or Hotmail account. Launch the Mail app from the Start menu.

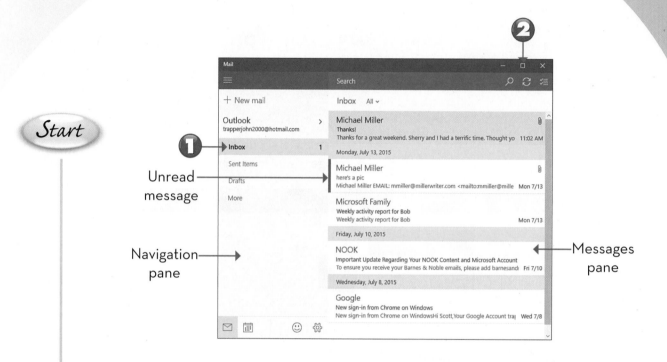

Unread message

Navigation pane

Messages pane

Start

When the Mail app launches, you see two panes within the window. The pane on the left is the navigation pane. Select **Inbox** from the navigation pane on the left to display a list of all your current messages in the Messages pane.

Click the **Maximize** button to maximize the Mail window and display a third pane on the right.

Continued

NOTE

Read and Unread The headers for unread messages are displayed with a blue line on the left. Messages you've read display normally. ■

TIP

Tile Info If you pin the Mail app to the Start menu, it becomes a "live" tile. Your most recent unread messages scroll across the face of the tile, and the number at the bottom left indicates how many unread messages you have. ■

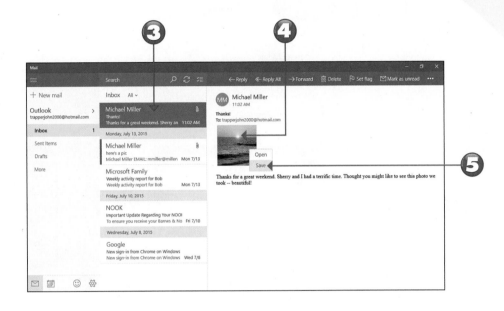

3 Click a message to view it in the content pane on the right. (If you haven't maximized the Mail window, the content of the selected message replaces the messages list.)

4 If a message includes a photograph, that photo's thumbnail image displays beneath the message text. (You might have to click the placeholder thumbnail to view the image.)

5 If the message has a photo or another file attached, right-click the item and click **Save** to download the file to your computer.

End

CAUTION

Beware Attached Viruses Beware of receiving unexpected email messages with file attachments. Opening the attachment could infect your computer with a virus or spyware! You should *never* open email attachments that you weren't expecting—or from senders you don't know. ∎

From:
To:
Cc:
Subject:

MOVING A MESSAGE TO ANOTHER FOLDER

New messages are stored in the Mail app's Inbox, which is actually a folder. Mail uses other folders too; there are folders for Drafts, Sent Items, Outbox (messages waiting to be sent), Junk (spam), Deleted Messages, and Stored Messages. For better organization, you can easily move messages from one folder to another.

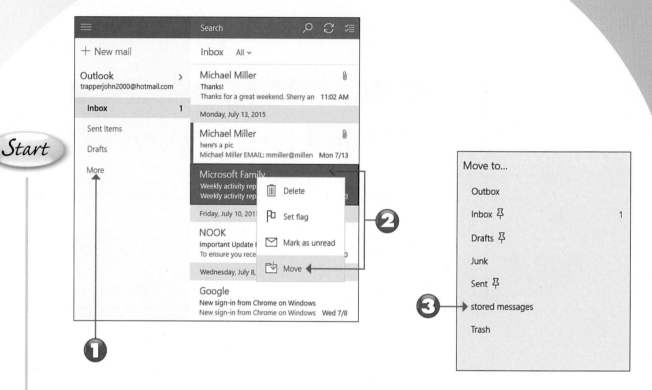

1 To view a list of folders, click **More**.

2 Right-click the message you want to move and click **Move**. This displays the Move To pane.

3 Click the destination folder.

End

NOTE

Drafts and Flags A *draft* message is one you've started but not yet sent. You *flag* a message (by clicking the Flag button in the options bar) when you want it to have increased importance. ∎

REPLYING TO AN EMAIL MESSAGE

Replying to an email message is as easy as clicking a button and typing your reply. The bottom of your reply "quotes" the text of the original message.

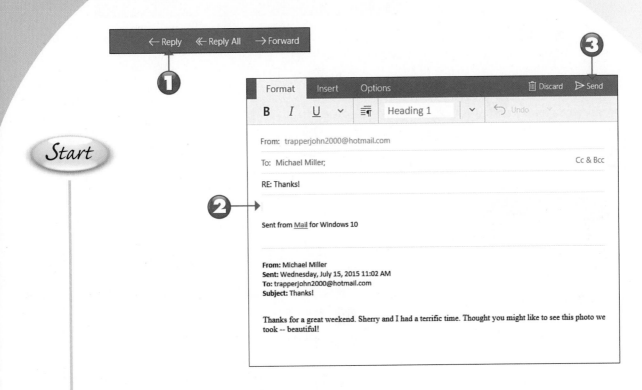

Start

① From an open message, click **Reply** at the top of the screen.

② Enter your reply at the top of the message; the bottom of the message "quotes" the original message.

③ Click the **Send** button when you're ready to send the message.

End

COMPOSING A NEW EMAIL MESSAGE

Composing a new message is similar to replying to a message. The big difference is that you have to manually enter the recipient's email address.

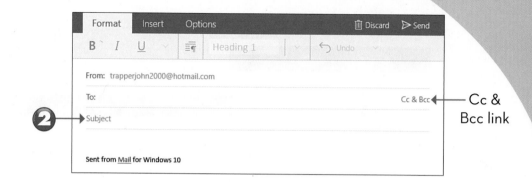

Cc & Bcc link

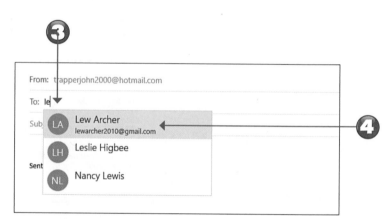

Start

1. Click **+ New Mail** at the top of the navigation pane to display the new message screen.

2. Click within the **Subject** area and type a subject for this message.

3. Click within the **To** box and begin entering the name or email address of the message's recipient.

4. Mail displays a list of matching names from your contact list; select the person you want to email.

Continued

TIP

Formatting Your Message When you're composing a message and you want to apply formatting to your text, click to select the Format tab. You can then use the Bold, Italic, Underline, and other buttons to format your message. ∎

TIP

Copying Other Recipients You can also send carbon copies (Cc) to additional recipients. (With a blind carbon copy, recipients cannot see the names of the Bcc recipients.) Click the **Cc & Bcc** link to display the Cc and Bcc boxes, and enter recipient names accordingly. ∎

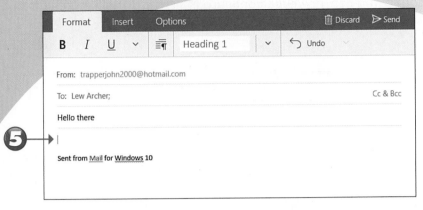

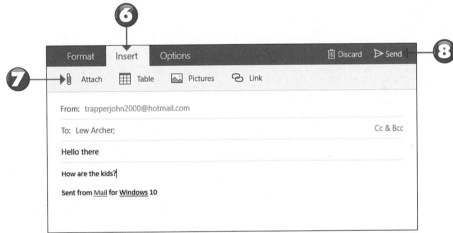

5 Click within the main body of the message area and type your message.

6 To attach a picture or another file to this message, click to select the **Insert** tab.

7 Click the **Attach** button to select the file you want to attach.

8 When you're ready to send the email, click the **Send** button at the top of the message.

End

TIP
Attaching Files One of the easiest ways to share a digital photo or another file with another user is via email, as an *attachment* to a standard email message. When the message is sent, the file travels along with it; when the message is received, the file is right there, waiting to be opened. ■

CAUTION
Large Files Be wary of sending extra-large files (2MB or more) over the Internet. They can take a long time to upload—and just as long for the recipient to download when received. ■

ADDING OTHER ACCOUNTS TO THE MAIL APP

By default, the Mail app sends and receives messages from the email account associated with your Microsoft account. You can, however, configure Mail to work with other email accounts, if you have them.

1

Start

2 → Accounts

3 → + Add account

1 From within the Mail app, click the **Settings** button (at the bottom of the navigation pane) to display the Settings pane.

2 Click **Accounts** to display the Accounts pane.

3 Click **+ Add Account** to display the Choose an Account window.

Continued

TIP

Switching Accounts To view the Inbox of another email account in the Mail app, click the right arrow next to the current account in the navigation pane and then select the other account. ■

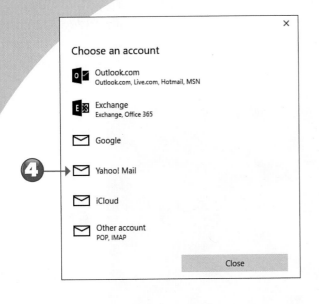

④ Click the type of account you want to add.

⑤ When the next screen appears, enter your email address and password.

⑥ Click the **Sign-In** button when done.

End

TIP

Account Types The Mail app lets you add Outlook.com, Exchange, Google (Gmail), Yahoo! Mail, iCloud, and other POP/IMAP email accounts. ■

MANAGING YOUR CONTACTS FROM THE PEOPLE APP

The people you email regularly are known as *contacts*. When someone is in your contacts list, it's easy to send her an email; all you have to do is pick her name from the list instead of entering her email address manually. All your Windows contacts are managed from the People app; this app connects to the Microsoft account you used to create your Windows account, so all the contacts from your main email account are automatically added. Launch the People app from the Start menu.

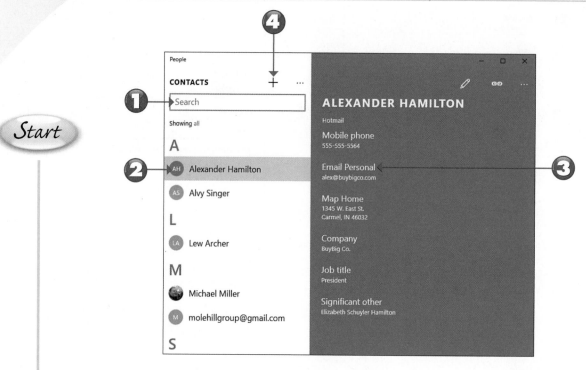

① To search for a specific contact, enter that person's name into the **Search** box then press **Enter**.

② Click or tap a person's name to view that person's contact information.

③ Click the person's email address to send this person an email in the Windows Mail app.

④ To add a new contact, click the **+** to display the New Contact screen.

Continued

NOTE

Social Contacts The People app centralizes all your contacts in one place, so you'll find not only email contacts but also Facebook friends and the people you follow on Twitter. So if a given person is a Facebook friend and is also in your email contact list, his Facebook information and his email address appear in the People app. ■

NOTE

First-Time Use The first time you launch the People app, you're prompted to add your Microsoft Account to the app. Do so, by entering your email address and password. You can later add other email accounts to the app. ■

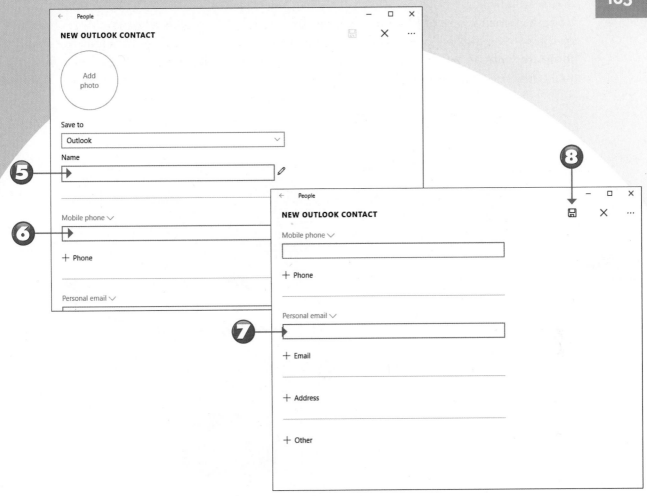

5 Enter the person's full name into the **Name** box.

6 Optionally, enter the person's mobile phone number into the **Mobile Phone** box.

7 Optionally, enter the person's email address into the **Personal Email** box.

8 Click **Save** when done.

End

NOTE

More Info To include additional email addresses, phone numbers, street addresses, or other information for this person, click **+ Email**, **+ Phone**, **+ Address**, or **+ Other** and enter the necessary information. ■

READING WEB-BASED EMAIL WITH GMAIL

Google's Gmail is one of the most popular free web-based email services. Anyone can sign up for a free Gmail account and then use any web browser to access his email from any computer with an Internet connection. Create a new Gmail account and access Gmail from mail.google.com.

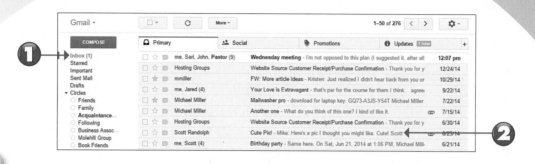

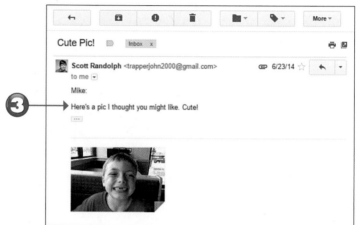

Start

 On Gmail's main page, click the **Inbox** link.

 Click the header for the message you want to view.

3 The selected message now appears; scroll down to read the entire text, if necessary.

End

NOTE

Other Web-Based Email Other popular web-based email services include Microsoft's Outlook.com (www.outlook.com) and Yahoo! Mail (mail.yahoo.com). ■

NOTE

POP/IMAP Versus Web-Based Email Most Internet service providers assign you an email account using the POP or IMAP protocols, which require the use of a separate email program, such as Microsoft Outlook. Web-based email doesn't require any new software programs, and you can use a web browser to access it from any computer or mobile device. ■

REPLYING TO A GMAIL MESSAGE

It's easy to reply to any message you receive. Just click **Reply** and then enter your new message!

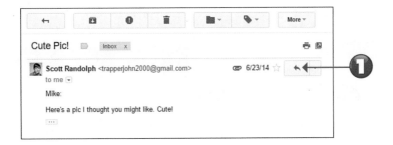

Start

1 From the open message, click **Reply**.

2 Enter your reply text in the message window.

3 Click the **Send** button to send your reply to the original sender.

End

NOTE

Quoted Text The text of the original message is automatically "quoted" at the bottom of the reply message. ■

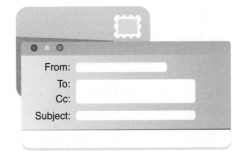

COMPOSING A NEW GMAIL MESSAGE

Composing a new message is similar to replying to a message, but you do it in a separate new message pane. The big difference is that you have to manually enter the recipient's email address.

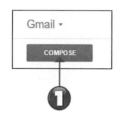

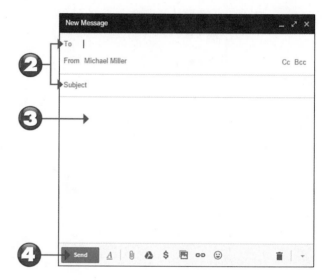

Start

1 Click the **Compose** button from any Gmail page to display the new message pane.

2 Enter the email address of the recipient(s) in the **To** box and then enter a subject in the **Subject** box.

3 Move your cursor to the main message area and type your message.

4 When your message is complete, send it to the recipient(s) by clicking the **Send** button.

End

TIP

Send to Multiple Recipients You can enter multiple addresses in the **To** box, as long as you separate the addresses with a semicolon, like this: books@molehillgroup.com; gjetson@sprockets.com. ■

TIP

Cc: and Bcc: Gmail also lets you send carbon copies (Cc:) and blind carbon copies (Bcc:) to additional recipients. (With a blind carbon copy, recipients of the email cannot see the names of the Bcc recipients.) Just click the **Cc** or **Bcc** links to add these addresses. ■

Like all other email programs and services, Gmail lets you attach photos and other types of files to your outgoing email messages.

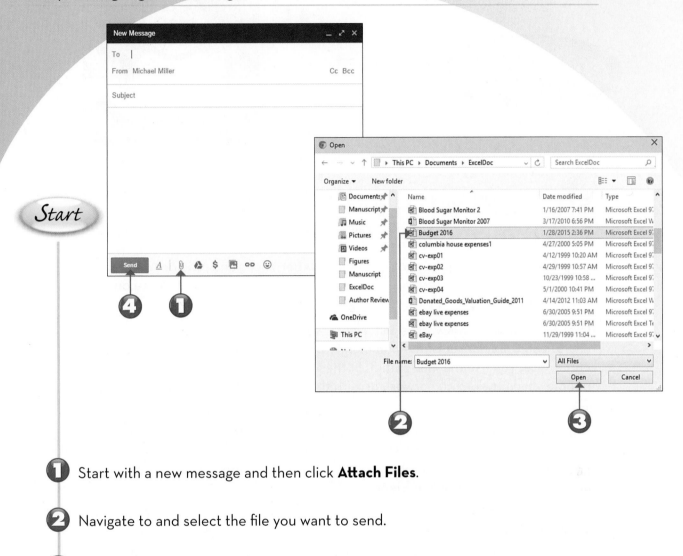

1 Start with a new message and then click **Attach Files**.

2 Navigate to and select the file you want to send.

3 Click **Open**.

4 Complete the message as normal and then send it by clicking the **Send** button.

End

SHARING WITH FACEBOOK AND OTHER SOCIAL NETWORKS

Social networking enables people to share experiences and opinions with each other via community-based websites. It's a great way to keep up-to-date on what your friends and family are doing.

In practice, a social network is just a large website that aims to create a community of users. Each user of the community posts his or her own personal profile on the site. You use the information in these profiles to connect with other people you know on the network or with those who share your interests.

The goal is to create a network of these online "friends," and then share your activities with them via a series of posts or status updates. All your online friends read your posts, as well as posts from other friends, in a continuously updated *News Feed*. The News Feed is the one place where you can read updates from all your online friends and family; it's where you find out what's really happening.

The biggest social network today is a site called Facebook; chances are all your friends are already using it. Other popular social networks include Pinterest and Twitter, both of which have their own unique characteristics.

COMPARING FACEBOOK, PINTEREST, AND TWITTER

Facebook

Pinterest

Twitter

FINDING FACEBOOK FRIENDS

Facebook (www.facebook.com) is the number one social network today, with more than 1 billion active users worldwide. After you've signed up, you can use Facebook to track down and keep in touch with all your friends and family—including old schoolmates and co-workers. You can then invite any of these people to be your Facebook friend; if they accept, they're added to your Facebook friends list.

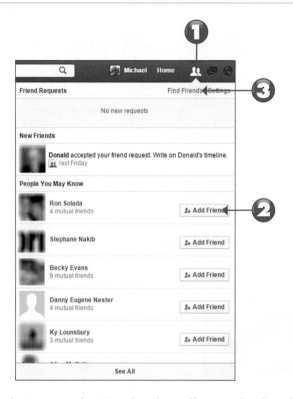

Start

Click the **Friend Requests** button on the Facebook toolbar to display the drop-down menu, which lists any friend requests you've received and offers a number of friend suggestions from Facebook in the People You May Know section.

Click the **Add Friend** button next to a person's name to add that person to your friends list.

To search for more friends, click **Find Friends** at the top of the menu to display your Friends page.

Continued

NOTE

Signing Up A Facebook account is free. Sign up at www.facebook.com. ∎

NOTE

Suggested Friends Facebook automatically suggests friends based on your personal history (where you've lived, worked, or gone to school), people you might know (friends of people you're already friends with), and Facebook users who are in your email contacts lists. ∎

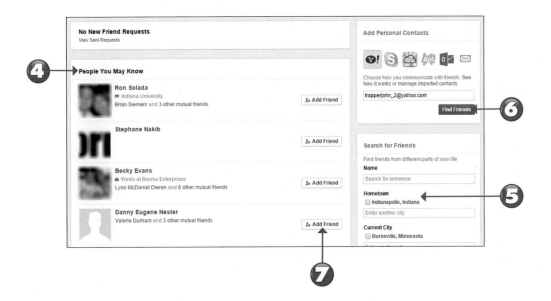

4 Scroll down the page to view other suggested friends from Facebook in the People You May Know section.

5 Filter the friend suggestions by making selections in the Hometown, Current City, High School, College or University, Employer, and Graduate School sections.

6 In the Add Personal Contacts section, click **Find Friends** to find people in your email contacts lists who are also on Facebook.

7 Click the **Add Friend** button for any person that you'd like to have on your friends list.

End

NOTE

Invitations Facebook doesn't automatically add a person to your friends list. Instead, that person receives an invitation to be your friend and can accept or reject the invitation. ■

TIP

Accepting a Friend Request To accept or reject any friend requests you've received, click the **Friend Requests** button on the Facebook toolbar. ■

READING THE NEWS FEED

Facebook's News Feed, found on your Facebook home page, is where you keep abreast of what all your friends are up to. When a person posts a *status update* to Facebook, it appears in your personal News Feed.

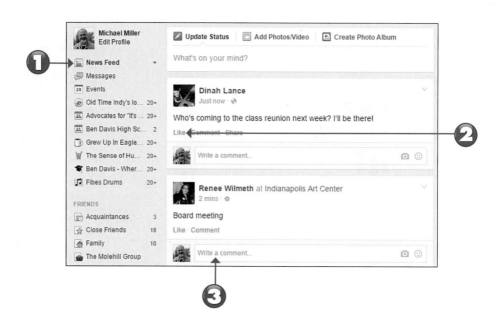

Start

From the sidebar menu, click the **News Feed** icon to open your home page with the News Feed displayed.

To "like" a status update, click **Like** under the post.

To comment on a status update, type your comment into the Write a Comment Box.

Continued

TIP

Top Stories By default, Facebook sorts the posts in your News Feed by importance—what Facebook calls your Top Stories. The problem is that Facebook's idea of what's important might not be what you find most important. To display *all* posts instead, click the **down arrow** next to News Feed in the sidebar menu and click **Most Recent**. ■

4 If a status update includes a link to a web page, click that link to open that page.

5 If a status update includes one or more photos, click a photo to view it in its own *lightbox*—a special window displayed on top of the News Feed.

6 If a status update includes a video, it might begin playback automatically. If not, click the **Play** arrow to play the video.

End

TIP
Sidebar Menu Use the sidebar menu on the left side of the home page to jump to different parts of the Facebook site. ■

TIP
Share an Update If you'd like to share a friend's post with your own friends, click **Share** under the post and then click **Share** in the pop-up menu to display the Share This Status dialog box. Click the **Share** button and select **On Your Own Timeline**, enter any comments you might have on this post into the Write Something box, and then click the **Share Status** button. ■

POSTING A STATUS UPDATE

The easiest way to let people know what's what is to post what Facebook calls a *status update*. Every status update you make is broadcast to everyone on your friends list, displayed in the News Feed on their home pages. A basic status update is text only, but you can also include photos, videos, and links to other web pages in your posts.

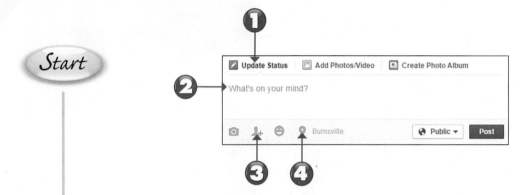

Start

1 On the Facebook home page, go to the Publisher box (labeled What's On Your Mind?) at the top of the page and click **Update Status**. (It's probably selected by default.)

2 Type your message into the What's on Your Mind box. As you do this, the box expands slightly.

3 If you're with someone else and want to mention that person in the post, click the **Tag People in Your Post** button and enter that person's name.

4 If you want to include your current location in your post, click the **Add a Location to Post** button and enter the city or place where you're located.

Continued

CAUTION

Social Networking Safety Make sure that you and your kids don't post overly personal information or incriminating photographs on Facebook or other social networks; you could attract online stalkers. Similarly, don't broadcast your every move on your profile page—and don't automatically accept friend requests from people you don't know. ■

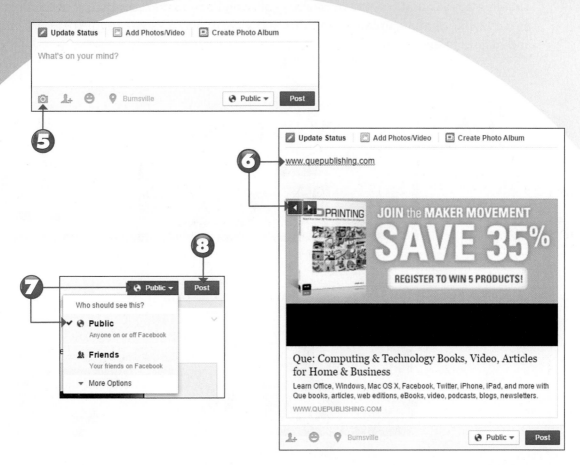

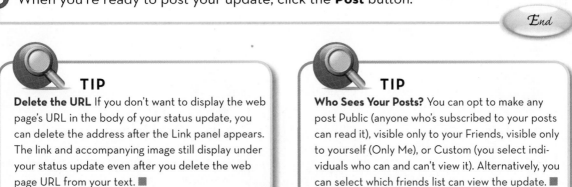

5 To include a picture or video with your post, click the **Add Photos to Your Post** (camera) button, and then select the photos or video you want.

6 To include a link to another web page, simply enter that page's URL in your status update. Facebook should recognize the link and display an image from the web page. If there is more than one image, click the right or left arrow buttons to select the one you want.

7 To determine who can read this post, click privacy button and make a selection.

8 When you're ready to post your update, click the **Post** button.

End

TIP
Delete the URL If you don't want to display the web page's URL in the body of your status update, you can delete the address after the Link panel appears. The link and accompanying image still display under your status update even after you delete the web page URL from your text. ■

TIP
Who Sees Your Posts? You can opt to make any post Public (anyone who's subscribed to your posts can read it), visible only to your Friends, visible only to yourself (Only Me), or Custom (you select individuals who can and can't view it). Alternatively, you can select which friends list can view the update. ■

VIEWING A FRIEND'S TIMELINE

You can easily check up on what a friend is up to by visiting that person's Timeline page. A Timeline page is that friend's personal profile on Facebook; it contains all of that person's personal information, uploaded photos and videos, and a "timeline" of that person's posts and major life events.

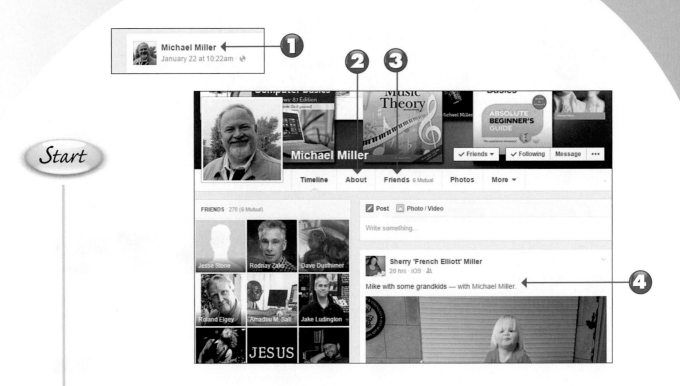

Start

① Click a person's name anywhere on the Facebook site to display his or her profile or Timeline page.

② View this person's full personal profile by clicking **About**.

③ View a list of this person's friends by clicking **Friends**.

④ View a person's status updates in reverse chronological order (newest first) on the Timeline.

End

TIP

Read the Book Learn more about Facebook in my book *My Facebook for Seniors*, 2nd Edition (Que, 2014). ■

TIP

Posting on a Friend's Page You can post a message on your friend's profile page by entering your text into the Write Something box near the top of the Timeline. ■

You can personalize your own profile page in a number of ways. You can change your profile picture, edit your personal information, and add and delete items to and from your Timeline.

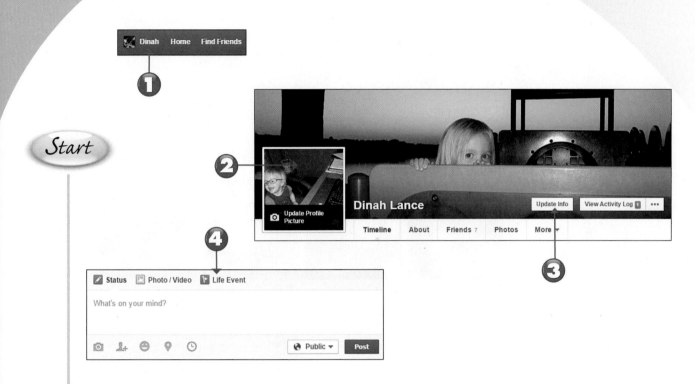

Start

1 Click your name in the Facebook toolbar or your picture in the sidebar menu to display your Timeline page.

2 To change your profile picture, click the picture and select **Update Profile Picture**.

3 To change your personal information, click the **Update Info** button.

4 To add a "life event" (something major in your life, such as getting a new job or getting married), go to the Publisher box, click **Life Event**, and choose the kind of life event you're adding.

End

TIP

Cover Image To add a cover image (banner) to the top of your Timeline page, mouse over the image area and click the **Add a Cover** button. To change an existing cover image, click the **Update Cover Photo** button. ■

TIP

Hide a Status Update If you've posted a status update that you'd rather not have visible, click the **down arrow** at the top-right corner of that item. Click **Hide from Timeline** to hide (but not delete) the status update. Click **Delete** to permanently remove the update from Facebook. ■

VIEWING A FRIEND'S PHOTOS

Facebook is a social network, and one of the ways we connect socially is through pictures. Facebook lets any user upload and store photos in virtual photo albums. It's easy, then, to view a friend's photos on the Facebook site.

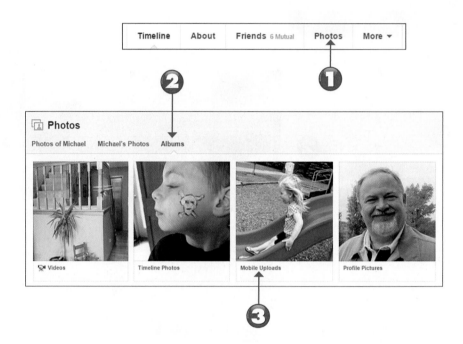

Start

1. Go to your friend's Timeline page and click **Photos** to display his Photos page.

2. Click **Albums** to display all of your friend's photo albums.

3. Click to open the desired photo album.

Continued

NOTE

Other Photos You can also select, from the top of the Photos page, to view Photos of This Person (all photos in which this person appears) or This Person's Photos (all the photos this person has uploaded, not organized into albums). ■

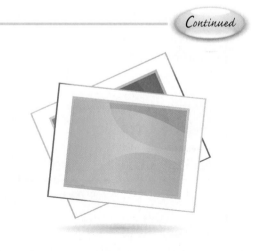

Mobile Uploads
Updated about 2 weeks ago

(4) Click the thumbnail of the picture you want to view.

(5) Go to the next picture by mousing over the current picture to display the navigational arrows, and then clicking the **right arrow**.

(6) Click the **X** to close the photo viewer.

End

TIP

Commenting and Liking To comment on the current picture, enter your message into the comments box. To like a photo, click **Like**. ■

TIP

Downloading a Picture To download the current picture to your own computer, mouse over the photo, click **Options**, and then click **Download** from the pop-up menu. ■

SHARING YOUR PHOTOS ON FACEBOOK

Facebook is a great place to share your personal photos with family and friends. You can upload new photos to an existing photo album or create a new album for newly uploaded photos.

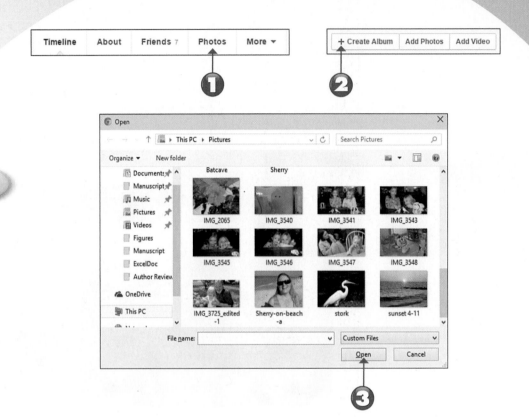

Start

 Open your Timeline page and click **Photos** to display your Photos page.

 Click the **Create Album** button to display the Select File(s) to Upload or the Open dialog box.

3 Select the photos you want to upload, and then click the **Open** button to see the Untitled Album page.

Continued

NOTE

Photo Requirements Facebook accepts photos in all popular file types, including JPG, PNG, GIF, TIFF, and BMP. Your picture files have to be no larger than 15MB in size and can't contain any adult or offensive content. You're also limited to uploading your own photos—that is, you can't copy and then upload photos from another person's website. ■

TIP

Uploading to an Existing Photo Album You can also upload photos to an existing photo album. From your Photos page, click **Albums** to display your existing photo albums, and then click to open the desired album. Click the **Add Photos** button to select new photos to upload to this album. ■

4 Enter information about this new album into the Untitled Album, Say Something About This Album, Where Were These Taken, and other boxes.

5 Enter information about each new photo beneath each photo.

6 If there are people in the photo you've uploaded, Facebook displays the album page with boxes around the faces. To "tag" that person in Facebook, click a face and then enter that person's name.

7 Click the **Post** button when ready.

End

TIP

High-Quality Photos To upload photos at their original resolution, check the **High Quality** option. This enables your friends to download your pictures at an acceptable resolution for printing. ■

TIP

Photo Privacy To determine who can view the photos in this album, click the **Privacy** button and make a selection—Public, Friends, Only Me, or Custom. ■

SHARING INTERESTING IMAGES WITH PINTEREST

Pinterest (www.pinterest.com) is a social network with particular appeal to women and people who like do-it-yourself projects. Unlike Facebook, which lets you post text-based status updates, Pinterest is all about images. The site consists of a collection of virtual online "pinboards" that people use to share pictures they find interesting. Users "pin" photos and other images to their personal message boards, and then share their pins with online friends.

Pinned image

Start

User who pinned the image

Board where image was pinned

Click the Pinterest logo to return to the home page.

Click the button with your name to view recent activity regarding your pins—typically, people who've repinned or liked your pins.

Click a pin to view it in more detail.

End

NOTE

What's in a Name? Pinterest is all about pinning items of interest—hence the name, a combination of *pin* and *interest*. ∎

NOTE

Pinterest Is Popular Pinterest boasts more than 70 million users. The average Pinterest user is a rural female, age 30 to 49, with a college degree and household income in the $50,000 to $75,000 range. ∎

When you find someone who posts a lot of things you're interested in, you can follow that person on Pinterest. Following a person means that all that person's new pins will display on your Pinterest home page.

Start

End

1. When you find a pin you like, click the name of the user who pinned it.

2. You now see that user's personal page. Click the **Follow** button to follow this user.

3. To follow a single board, instead of all of a user's pins, click the **Follow** button for that board.

TIP

Popular Categories The most popular categories on Pinterest are Home, Arts and Crafts, Style and Fashion, and Food. Of these, Food is the category most likely to be repinned. ■

FINDING AND REPINNING INTERESTING PINS

Some people say that Pinterest is a little like a refrigerator covered with magnets holding up tons of photos and drawings. You can find lots of interesting items pinned from other users—and then "repin" them to your own personal pinboards.

Start

1 Enter the name of something you're interested in into the Search box at the top of any Pinterest page, and then press **Enter**.

2 Pinterest now displays pins that match your query. Mouse over the item you want to repin and click the **Pin It** button.

Continued

NOTE

Repins About 80% of the pins on Pinterest are actually repins. ■

TIP

Searching for Boards and People To search for boards instead of pins, click **Boards** at the top of the search results page. To display Pinterest users who match your query, click **Pinners**. ■

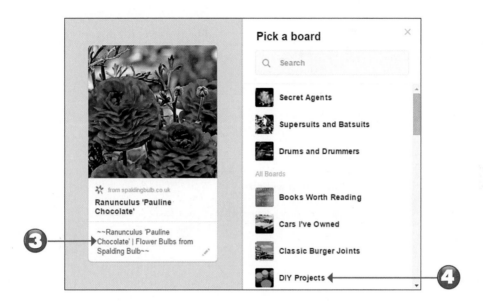

3 When the Pick a Board panel appears, accept the previous user's description or add your own to the Description box.

4 Scroll through the board list and select which board you want to pin this item to. The item is now pinned to that board.

End

TIP

Keep or Replace You can keep the original pinner's description or replace it with a new description of your own. If you want to truly personalize your pins, it's best to use your own descriptions, even when you repin. ■

PINNING FROM A WEB PAGE

You can also pin images you find on nearly any web page. It's as easy as copying and pasting the page's web address.

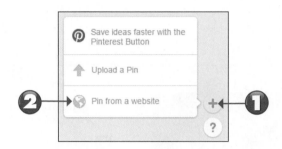

Start

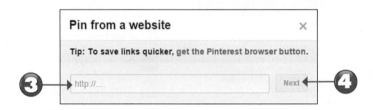

1 Click the **+** button at the bottom right of any Pinterest page to display the menu of options.

2 Click **Pin from a Website** to display the Pin from a Website panel.

3 Enter the web address (URL) of the page you want to pin into the text box.

4 Click the **Next** button.

Continued

TIP
Pinnable Images When you're looking for items to pin, consider the image on a web page. Look for images that look good at thumbnail size and will be appealing to other users. ■

TIP
Easier Pinning Make pinning easier by installing a Pinterest button in your web browser. Learn more at about.pinterest.com/goodies/#browser. ■

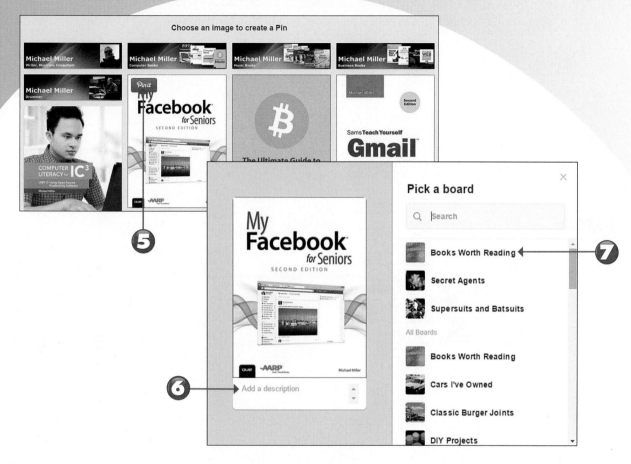

5 Pinterest now displays a page containing all the images found on the selected web page. Click the **Pin It** button for the image you want to pin.

6 When the Pick a Board panel appears, enter a short (500 characters or fewer) text description of or comment on this image into the **Description** box.

7 Scroll down the board list and select the board to which you want to pin this image. The image is now pinned to this board.

End

TIP

Description Although a text description is optional, it's always a good idea to describe or comment on the item you're pinning. If you don't enter a description, people won't be able to find your pin by searching. ■

TIP

Read the Book Learn more about Pinterest in my book *My Pinterest* (Que, 2012). ■

TWEETING WITH TWITTER

Twitter is a *microblogging* service that lets you create short (up to 140 characters in length) text posts that keep your friends and family informed of your latest activities. Anyone subscribing to your posts receives updates via the Twitter site.

Start

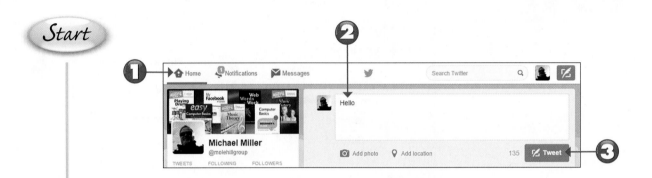

1. Go to Twitter's home page (www.twitter.com) and click the **Home** icon in the toolbar.

2. Enter up to 140 characters into the **What's Happening** box.

3. Click the **Tweet** button.

End

NOTE

Microblogging A *blog* (short for *web log*) is a means of sharing personal observations over the Internet, kind of like a web-based diary. Whereas a normal blog post can be of any length, a *microblog* is limited to very short messages—like the text message you send via mobile phone. ∎

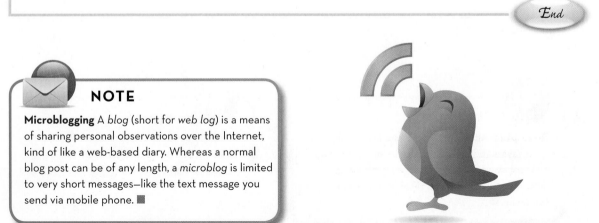

Twitter lets you "follow" what other users are doing on Twitter. After you've registered and signed in, the Twitter home page displays "tweets" from users you've decided to follow.

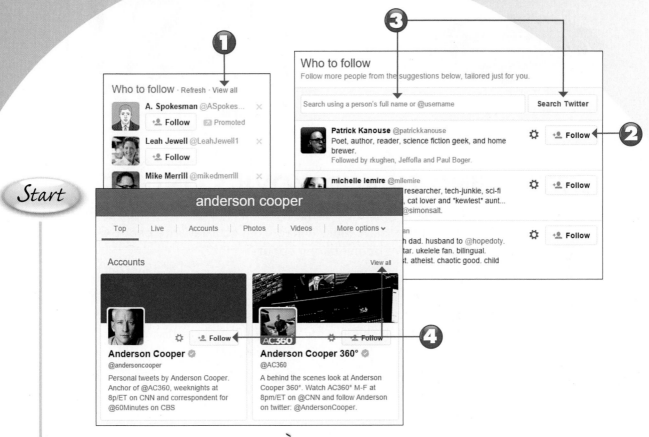

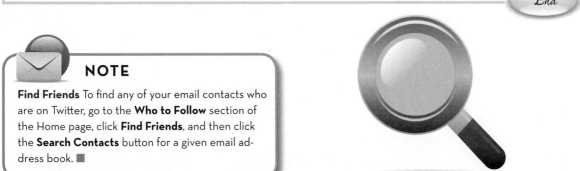

① From Twitter's Home page, go to the Who to Follow section and click **View All**.

② Twitter suggests people and accounts to follow. To follow a given person, click the **Follow** button.

③ To search for specific people on Twitter, enter that person's full name or Twitter username into the search box and then click **Search Twitter**.

④ Twitter displays users who match your query. Click **View All** to view more people; click the **Follow** button to follow a user.

NOTE

Find Friends To find any of your email contacts who are on Twitter, go to the **Who to Follow** section of the Home page, click **Find Friends**, and then click the **Search Contacts** button for a given email address book. ∎

Chapter 13

WATCHING TV AND MOVIES ONLINE

Want to rewatch last night's episode of *The Voice*? Or the entire season of *Big Bang Theory*? How about a classic music video from your favorite band? Or that latest "viral video" you've been hearing about?

Here's the latest hot thing on the Web: watching your favorite television shows, movies, and videos online, via your web browser. If you have a fast enough Internet connection, you can find tens of thousands of free and paid videos to watch at dozens of websites, including YouTube, Hulu, and Netflix. You can even purchase and download videos to your PC—and watch them anytime, at your convenience.

PLAYING MOVIES WITH THE WINDOWS MOVIES AND TV APP

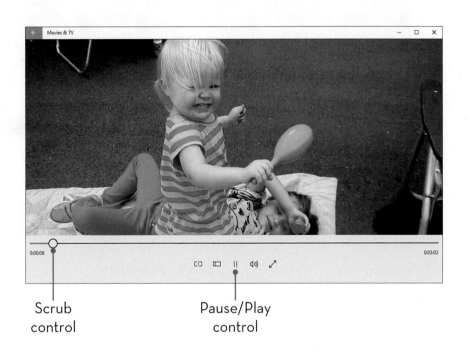

Scrub control

Pause/Play control

WATCHING MOVIES ON NETFLIX

Arguably the most popular site for streaming movies is Netflix (www.netflix.com), which offers a mix of both classic and newer movies, classic television shows, and original programming. You can watch Netflix in your web browser or via the Netflix app for Windows.

Start

1. Launch the Netflix app from the Start menu. The first section you see displays videos you haven't yet finished watching (Continue Watching). Scroll right to view other sections.

2. Click the title of any section to view more options of that type.

3. Click the **Options** button to view the available genres.

Continued

NOTE

Netflix App Download the (free) Netflix app from the Windows Store. After it's installed, you can create a new Netflix account or log in with an email address and password from an existing account. ■

TIP

User Profiles The Netflix website lets you create separate user profiles for up to five viewers in your household; each profile includes recommendations specific to that viewer. To change viewers from within the Netflix app, click the user profile at the top-right corner and make a different selection. ■

4 Click a genre to view movies of that type.

Continued

TIP

Subscription Fees Netflix isn't free. You pay $8.99/ month for unlimited streaming video online. ■

TIP

Netflix on the Web You can also access Netflix directly from any web browser. Just go to www.netflix. com and sign up or sign on. ■

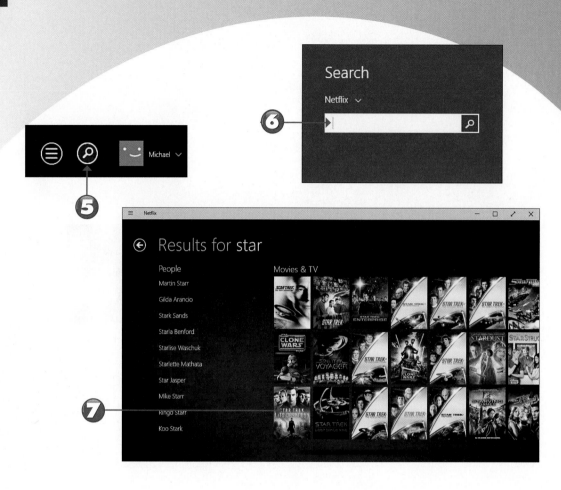

 To search for a specific movie or show, click the **Search** (magnifying glass) icon to display the search pane.

Enter the name of the movie or show into the Search box and then press **Enter**.

When you find a movie or show you want to watch, click it. The detail page for that movie or show displays.

Continued

NOTE

Streaming Video Most movies and TV shows you watch online are not downloaded to your computer. Rather, they flow in real time from the host website to your PC, over the Internet, using a technology called *streaming video*. With streaming video, programs can start playing back almost immediately, with no time-consuming downloading necessary. ■

NOTE

DVD Rental Netflix also offers a separate DVD-by-mail rental service, with a separate subscription fee. ■

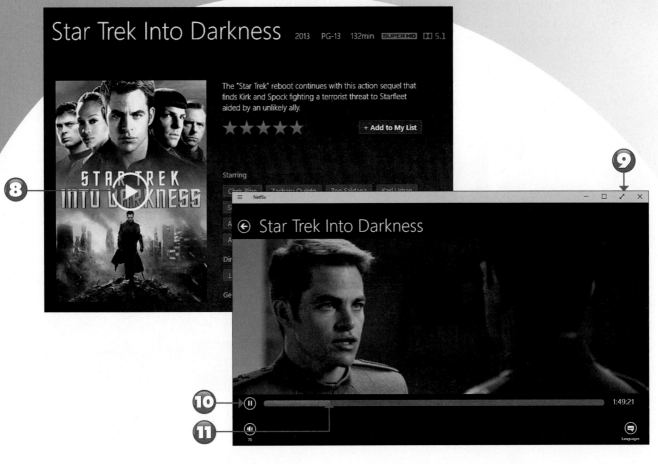

8 Click the **Play** button on the image to watch the program.

9 Netflix begins playing the movie or show you selected. Click the **Full-screen** button at the top of the window to display the movie full-screen.

10 Right-click anywhere on the screen to display the options bar. Click the **Pause** button to pause playback; the Pause button changes to a Play button. Click the **Play** button to resume playback.

11 Click and drag the scrub (slider) control to move directly to another part of the movie.

End

TIP

TV Shows If you choose to watch a TV show, you can usually choose from different episodes in different seasons. Select a season to see all episodes from that season and then click the episode you want to watch. ■

WATCHING TV SHOWS ON HULU

If Netflix is the best website for movies, Hulu is the best site for television programming. Hulu offers episodes from major-network TV shows, as well as new and classic feature films, for online viewing. The standard free membership offers access to a limited number of videos; a full membership is $7.99/month for a larger selection of newer shows.

Start

1 From your web browser, go to www.hulu.com. Hulu's home page displays a series of featured programs. Scroll down to view recommended programming by type.

2 To view television programs by genre, click **TV** at the top of the page.

Continued

NOTE

Hulu App There is also a Hulu app available (free) from the Windows Store. Unfortunately, this Universal app works only with a full Hulu subscription ($7.99/month). When you use the Hulu website, you can access both the free and the subscription service. ■

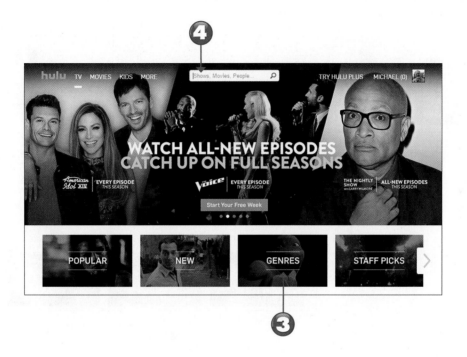

 To view programs by genre, click the **Genres** tile.

 To search for a specific show, enter the name of the show into the top-of-page Search box and then press **Enter**.

Continued

TIP

Movies Hulu also offers a variety of movies for online viewing. The standard free membership has a very limited selection of movie programming, typically documentaries and movie trailers. The $7.99 Hulu Plus membership offers a much larger selection of movies. ■

TIP

Network Websites Most major broadcast and cable TV networks offer their shows for viewing from their websites. These include ABC (abc.go.com), CBS (www.cbs.com), Comedy Central (www.comedycentral.com), CW (www.cwtv.com), Fox (www.fox.com), NBC (www.nbc.com), Nick (www.nick.com), Showtime (www.sho.com), TNT (www.tntdrama.com), and USA Network (www.usanetwork.com). ■

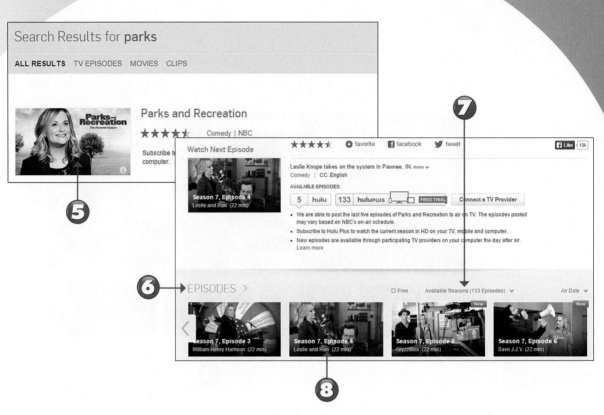

5 Click the tile for the show you want to watch.

6 When the detailed program page appears, scroll down to view available episodes, clips, and extras.

7 To view seasons by episode, click **Available Seasons** and then click the season you want.

8 Click the tile for the episode you want to watch.

Continued

(9) Hulu begins playing the program you selected. Move your mouse over the screen to display the playback controls.

(10) Click the **Pause** button to pause playback; the Pause button changes to a Play button. Click the **Play** button to resume playback.

(11) Click and drag the scrub (slider) control to move directly to another part of the program.

(12) Click the **Full-screen** button to view the program full-screen on your computer display.

End

WATCHING VIDEOS ON YOUTUBE

The most popular video site on the Web is YouTube. This site is a video-sharing community; users can upload their own videos and watch videos uploaded by other members. YouTube also offers a variety of commercial movies, TV shows, and music videos.

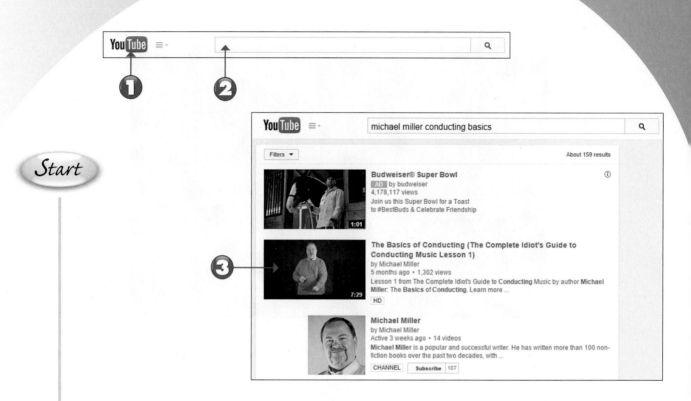

Start

① From your web browser, enter **www.youtube.com** into the Address bar, and then press **Enter**. You now see YouTube's home page.

② To search for a video, enter one or more keywords that describe the type of video you're looking for into the Search box, and then press **Enter** or click the **Search** (magnifying glass) button.

③ When the list of matching videos appears, click the video you want to watch.

Continued

TIP

Movies In addition to its user-uploaded videos, YouTube offers commercial movies. Some movies are free; others can be rented on a 48-hour pass for as low as $1.99. ■

TIP

Playlists To add a video to a playlist, click the **Add To** button under the video player, and then click a playlist. You can view your playlists by clicking the **Menu** button next to the YouTube logo (at the top of the page); click a playlist to begin playback. ■

The Basics of Conducting (The Complete Idiot's Guide to Conducting Music Lesson 1)

Michael Miller

Subscribe 107

1,305

+ Add to   < Share   ••• More

👍 2 👎 0

4 When the video page appears, the video begins playing automatically.

5 Click the **Pause** button to pause playback; click the button again to resume playback.

6 Click the **Full-screen** button to view the video on your entire computer screen.

7 Click the **thumbs-up** button to "like" the video.

End

TIP

Sharing Videos Find a video you think a friend would like? Click the **Share** button under the video player. You can then opt to email a link to the video, share the video on Facebook, or tweet a link to the video on Twitter. ∎

TIP

Uploading Videos to YouTube To upload your own home movies to YouTube, click **Upload** at the top of any page. On the next screen, click **Select Files from Your Computer**, and then navigate to and select the video you want to upload. After the video is uploaded, you can add a title and description and choose a thumbnail image for the video. ∎

PURCHASING AND DOWNLOADING MOVIES AND TV SHOWS FROM THE ITUNES STORE

One downside to Netflix, Hulu, and other streaming video services is that they don't offer more current theatrical movies. When you want to watch a movie fresh out of the theaters, you have to purchase or rent it, instead—which you can do online via Apple's iTunes Store.

TV shows

Movies

Start

Continued

1 Launch the iTunes app from the Start menu, and then click **iTunes Store** at the top of the window.

2 From within the Store, click either the **Movies** or **TV Shows** icon in the top-left corner.

3 Navigate to and click the movie or show you want to purchase.

NOTE

Install the iTunes App To access the iTunes Store, you first need to install the iTunes software on your computer. You can do this, free, from www.apple.com/itunes/. ∎

TIP

Other Online Video Stores You can purchase movies and TV shows from several other sites online. These include Amazon Instant Video (www.amazon.com/primeinstantvideo), CinemaNow (www.cinemanow.com), and Vudu (www.vudu.com). ∎

④ If a program is available for purchase, you see a Buy button with the price listed. Click **Buy** to purchase and download the item to your PC.

⑤ If a program is available for rental, you see a Rent button with the price listed. Click **Rent** to download the movie for limited-time viewing.

End

TIP

Apple Account Before you can purchase items from the iTunes Store, you have to create an Apple account and enter your credit card information. You might be prompted to do this the first time you click to purchase, or you can create your account manually, at any time, by clicking the **Sign In** button at the top right of the iTunes window and, when prompted, clicking the **Create New Account** button. ∎

NOTE

Purchase Versus Rent When you purchase a movie or TV show, it downloads to your computer and you're free to watch it as often and as long as you like; it's yours, you bought it. When you rent a program, however, it's available to you only for a limited time. With most rentals, you have 30 days to start watching the program, and then must finish watching it within 24 hours of first clicking the Play button. ∎

WATCHING VIDEOS WITH ITUNES

The iTunes app isn't just for browsing Apple's iTunes Store. You also use iTunes to watch all the movies and TV shows you download.

Start

1. Launch the iTunes app from the Start menu, open the iTunes Store, and then select either **Movies** or **TV Shows**.

2. Click **My Movies** or **My TV Shows** at the top of the window.

3. Double-click the program you want to watch.

Continued

4 When the video begins to play, move your mouse to display the playback controls.

5 Click the **Pause** button to pause playback; click the button again to resume playback.

6 Click the **Full-screen** button to view the video on your entire computer screen. Press **Esc** to return to the normal iTunes window.

End

WATCHING DIGITAL VIDEOS WITH THE MOVIES AND TV APP

Windows comes with its own app for viewing digital videos stored on your PC. You can use the Movies and TV app for viewing home movies you've created or videos you've downloaded from elsewhere.

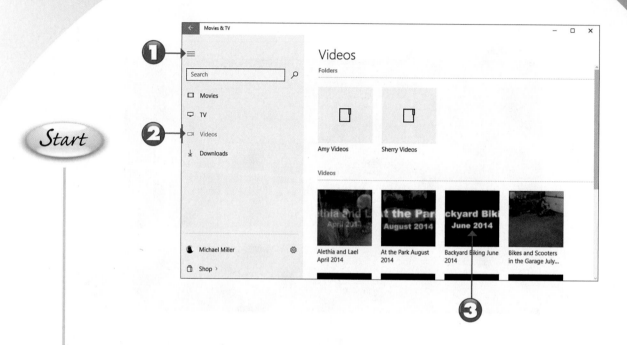

Launch the Movies and TV app from the Start menu, and then click the **Menu** button to expand the navigation sidebar. (You might also want to expand the window to make it larger on your desktop.)

Click **Videos** in the navigation sidebar to view those videos stored on your PC.

Click the video you want to watch.

Continued

NOTE

Movies and TV Stores Microsoft offers its own Movies Store and TV Store for purchasing and renting movies and TV shows online. You access these stores from the Movies and TV app; just click **Shop** in the navigation sidebar. ■

4 Mouse over the video to display the playback controls.

5 Click the **Pause** button to pause playback; click the button again to resume playback.

6 Click and drag the scrub (slider) control to move to a specific point within the video.

7 Click the **Full-screen** button to display the video full-screen.

End

Chapter 14

PLAYING DIGITAL MUSIC

Your personal computer can do more than just compute. It can also serve as a fully functional music player!

You can use your PC to listen to music played from Spotify, Pandora, and other streaming music services. You can download and play music you purchase from the iTunes Store and other online music stores. You can even use your PC to listen to music the old-fashioned way, from compact disc!

COMPARING STREAMING MUSIC SERVICES

Spotify

Pandora

Google Play
Music
All Access

iHeart
Radio

Rhapsody
Premier

STREAMING MUSIC ONLINE WITH SPOTIFY

If you're a music lover, you can listen to pretty much any song you like online, from one of a number of *streaming music services*. These services don't download music files to your computer; instead, music is streamed to you in real time, over the Internet. One of the most popular of these music services is Spotify (www.spotify.com), which offers both free and paid plans.

Start

1. Launch the Spotify app from the Windows Start menu, and then create a new account or sign into your existing account. (You can also log in with your Facebook account, if you have one.)

2. To browse by genre, click **Genres & Moods**.

3. Click a genre tile to view music of that type.

4. To search for a specific song, album, or artist, enter your query into the top-of-page Search box and then press **Enter**.

Continued

NOTE

Download the Spotify App Spotify offers a web-based version you can access via your web browser, or a standalone Spotify app that offers enhanced functionality. Access both at www.spotify.com. ■

NOTE

Free Versus Paid Spotify's basic membership is free, but you're subjected to commercials every few songs. If you want to get rid of the commercials (and get on-demand music on your mobile devices too), you need to pay for a $9.99/month subscription. ■

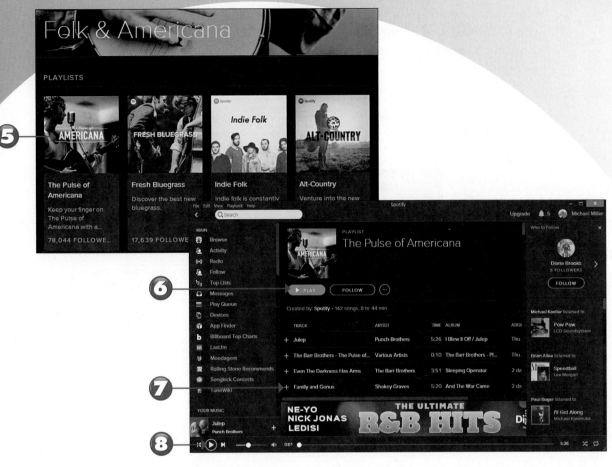

5 Click a playlist, an album, or an artist to view all included songs.

6 Click the green **Play** button to play all the songs in the playlist or album, or by that artist.

7 Double-click a song title to play that particular track.

8 Use the playback controls at the bottom left to pause, rewind, or fast-forward playback, or to raise or lower the volume.

End

NOTE

On-Demand Versus Personalized Services There are two primary types of delivery services for streaming audio over the Internet. The first model, typified by Spotify, lets you specify which songs you want to listen to; we call these *on-demand services*. The second model, typified by Pandora, is more like traditional radio in that you can't dial up specific tunes; you have to listen to whatever the service beams out, but in the form of personalized playlists or virtual radio stations. ■

NOTE

Other On-Demand Streaming Music Services In addition to Spotify, other on-demand streaming music services include Apple Music (www.apple.com/music/), Google Play Music All Access (play.google.com/about/music/), Rdio (www.rdio.com), Rhapsody Premier (www.rhapsody.com), and Slacker Premium (www.slacker.com). Most subscriptions run $9.99/month. ■

STREAMING MUSIC ONLINE WITH PANDORA

Pandora differs from Spotify in that you can't pick specific tracks to play. It's more like a traditional radio station, in that you listen to the songs Pandora selects for you, along with accompanying commercials. You create your own personalized stations, however, by choosing a song or an artist and then letting Pandora select other songs like the one you picked. You access Pandora from your web browser, at www.pandora.com.

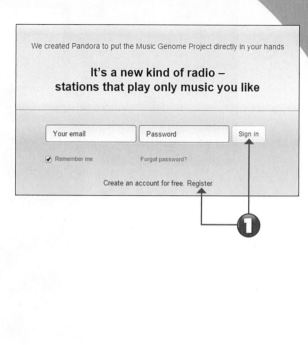

1 From the Pandora website (www.pandora.com) create a new account or log in to your existing account.

2 To create a new station, enter the name of a song, a genre, an artist, or a composer into the **Create Station** box at the top-left corner, and then press **Enter**.

3 The new station is added to your station list on the left side of the page. Click a station to begin playback; information about this track and artist is now displayed.

Continued

NOTE

Free Versus Paid Pandora's basic membership is free, but ad-supported. (You have to suffer through commercials.) To get rid of the commercials, pay for the $4.99/month Pandora One subscription. ■

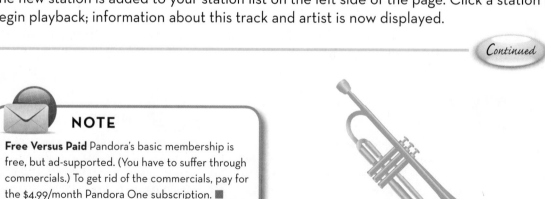

4 To pause playback, click the **Pause** button at the top of the page. Click **Play** to resume playback.

5 To like the current song, click the **thumbs up** button. Pandora will now play more songs like this one.

6 If you don't like the current song, click the **thumbs down** button. Pandora will now skip to the next song, not play the current song again, and play fewer songs like it.

7 To skip to the next song without disliking it, click the **next track** button.

End

NOTE

Other Personalized Radio Streaming Music Services In addition to Pandora, other personalized radio streaming music services include Apple's iTunes Radio (www.apple.com/itunes/itunes-radio/), Rhapsody unRadio (www.rhapsody.com), and Slacker Radio (www.slacker.com). Most offer both free (ad-supported) and paid services. ■

NOTE

Local Radio Stations Online If you'd rather just listen to your local AM or FM radio station—or to a radio station located in another city—you can do so over the Internet. Both iHeartRadio (www.iheart.com) and TuneIn (www.tunein.com) offer free access to local radio stations around the world. ■

DOWNLOADING MUSIC FROM THE ITUNES STORE

If you use an iPhone or iPad, chances are you download your music from Apple's iTunes Store, which has more than 37 million tracks available for downloading at prices ranging from 69¢ to $1.29 each. You play music purchased at the iTunes Store with Apple's iTunes music player application—which you also need in order to access the iTunes Store.

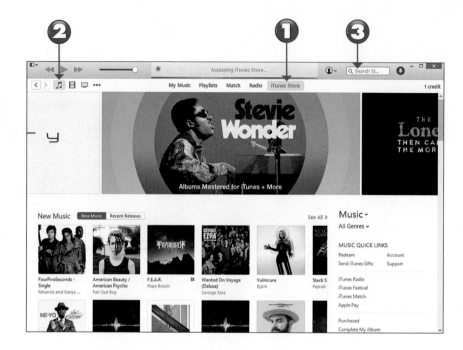

 Launch the iTunes app from the Start menu, and then click **iTunes Store** at the top of the window. This connects you to the Internet and displays the Store's home page.

 To view only music items, click the **Music** icon in the iTunes toolbar.

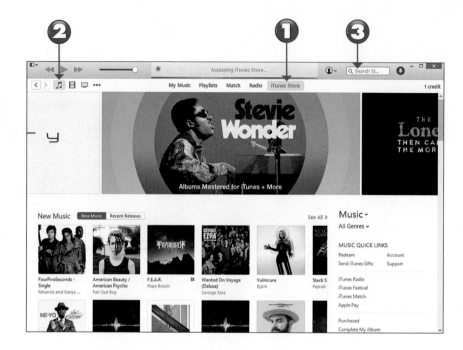 To search for something specific, enter the song title, album title, or artist name into the Search box at the top of the window and press **Enter**.

Continued

NOTE

iTunes You use the iTunes app to shop the iTunes Store, play the music you purchase, and manage the content of your iPhone, iPad, or iPod. Download the iTunes app free at www.apple.com/itunes/. ■

TIP

More in the Store The iTunes Store offers more than just music for download. iTunes also sells movies, TV shows, music videos, podcasts, audiobooks, and eBooks (in the ePub format). ■

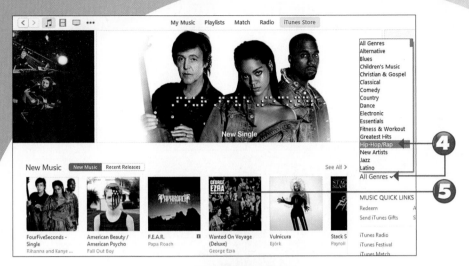

 To browse music by category, click **All Genres** on the right and select a category.

To view all the tracks in an album, click the album cover.

To purchase an individual track, click the **Buy** (price) button for that track.

To purchase an entire album, click the **Buy** button for that album.

End

TIP

Apple Account Before you can purchase items from the iTunes Store, you have to create an Apple account and enter your credit card information. You might be prompted to do this the first time you click to purchase, or you can create your account manually, at any time, by clicking the **Sign In** button at the top right of the iTunes window and, when prompted, clicking the **Create New Account** button. ■

TIP

Other Online Music Stores The iTunes Store is the store of choice if you have an Apple iPhone, iPod, or iPad; it fits in seamlessly with the Apple infrastructure and offers music in the AAC file format. If you prefer to get your digital music in the more universal MP3 file format, check out the Amazon Digital Music Store (www.amazon.com/mp3/) and Google Play Music store (play.google.com/store/music). ■

PLAYING DIGITAL MUSIC WITH ITUNES

You use the iTunes app to play all digital music that you've downloaded from the iTunes Store or that is otherwise stored on your computer.

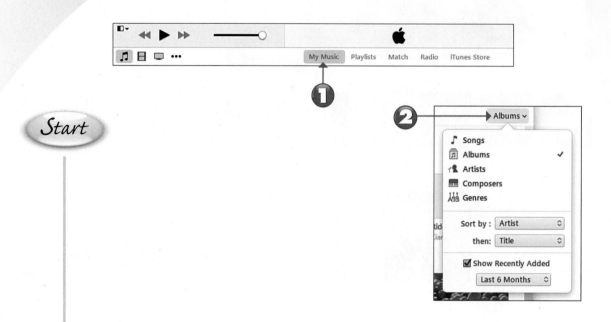

Start

1 Launch the iTunes app from the Start menu, and then click **My Music** at the top of the window.

2 To view all the tracks in your library, click the **down arrow** at the top-right corner and click **Songs**. To view your music organized by original album, click **Albums**. To view your music organized by artist, click **Artists**. To view your music organized by genre, click **Genres**.

Continued

Shuffle playback

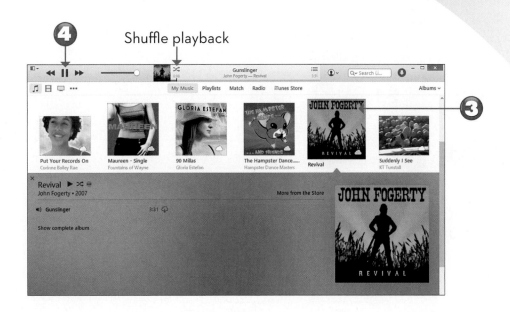

③ Navigate to the track, album, artist, or genre you want to play, and then double-click that item.

④ Click the **Pause** button to pause playback; click the **Play** button to resume playback.

End

TIP

Playlists You can put multiple songs together into a single *playlist* for future playback. To view and play back your playlists, click **Playlists** at the top of the iTunes window; a new Playlists pane opens on the left. To create a new playlist, click the + button at the bottom of the Playlists pane and then click **New Playlist**. To add songs to a playlist, drag individual tracks from the Content pane into the Playlists pane. ▪

TIP

Shuffle Playback To play an album or a playlist in random order, click the **Shuffle** icon in the mini-player at the top of the iTunes window. ▪

PLAYING A CD WITH ITUNES

If your computer includes a CD or CD/DVD drive, you can use it to play music CDs. The iTunes app doubles as a music player for your CDs.

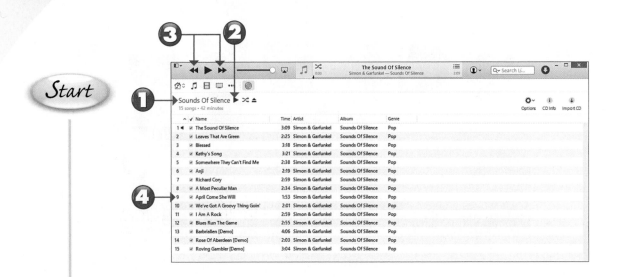

Start

1. Launch the iTunes app, and then insert the music CD into your PC's CD/DVD drive. The CD should automatically appear in the iTunes window, with all the tracks listed.

2. Click the **Play** button to begin playback; click **Pause** to pause playback.

3. Click the **Forward arrow** to skip to the next track on the CD. Click the **Back arrow** to skip to the previous track.

4. Double-click a specific track to jump to that track.

End

TIP

Ripping CDs The iTunes app can also import or "rip" CDs to digital files, stored on your computer. Just insert the CD and then click **Import CD** in the iTunes app. ∎

PURCHASING MUSIC FROM THE GROOVE MUSIC APP

Windows 10 includes the Groove Music app for both playing digital music and purchasing music online from the Windows Store.

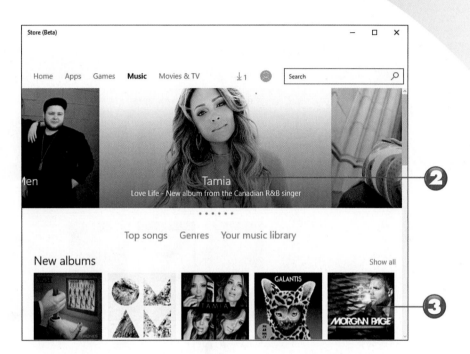

Start

1 Launch the Groove Music app from the Windows Start menu, and then click the **Shop for Music** icon to open the Windows Store app.

2 Scroll down to browse new albums, top-selling songs, top albums, top artists, and genres.

3 Click an item to view and purchase.

End

PLAYING DIGITAL MUSIC WITH THE GROOVE MUSIC APP

You can also use the Groove Music app to play digital music stored on your computer.

Start

1. Launch the Groove Music app from the Windows Start menu. To display all the albums in your collection, click **Albums**.

2. To display music organized by artist, click **Artists**.

3. To display individual tracks, click **Songs**.

4. Double-click an artist or album to view all tracks within.

Continued

5 Click the **Play** button to play all tracks within that artist or album.

6 Double-click a specific track to play only that track.

7 Click the **Pause** button to pause playback; click **Play** again to resume playback.

8 Click the **Turn Shuffle On** button to play tracks in a random order.

End

TIP
Search for Music To search for specific tunes or artists in your collection, click the **Search** button in the Navigation pane and enter the name of the song or artist you're looking for. ■

VIEWING AND EDITING DIGITAL PHOTOS

The traditional film camera is a thing of the past. These days, everybody uses a digital camera or smartphone camera—which you can easily connect to your PC. After it's connected, you can transfer all the photos you take to your computer's hard disk, view them on your computer monitor, share them with friends and family via Facebook and other social media, and even edit your pictures to make them look better.

The Photos app included with Windows 10 helps you find and view all the photos stored on your PC. It even lets you touch up your photos with easy-to-use editing functions.

VIEWING A PICTURE WITH THE PHOTOS APP

Automatic
enhance

Rotate
photo

Share   Play   Edit   Delete
photo slideshow   photo   photo

Options

TRANSFERRING PICTURES VIA USB

Whether you have a digital camera or a smartphone, you can easily transfer photos from that device to your PC, simply by connecting the two with the USB cable that came with your device. After they're connected, Windows sees your camera or smartphone as another drive on your system; it's then a simple matter to copy files from your device to your PC.

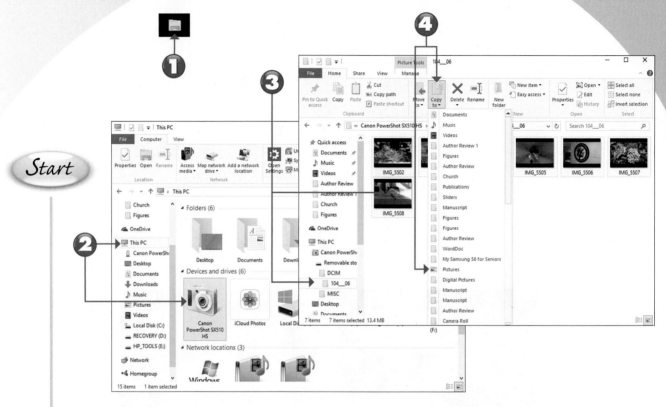

Start

End

1. With your camera or smartphone connected to your PC, click **File Explorer** on the taskbar to open File Explorer.

2. Click **This PC** in the Navigation pane, and then double-click the icon for your camera or smartphone.

3. Navigate to and open the folder where the photos reside (usually labeled DCIM), and then hold down the **Ctrl** key and click each photo you want to transfer.

4. Select the **Home** ribbon and click **Copy To**, and then select **Pictures**.

TIP

Automatic Action Windows might recognize when you connect your camera or smartphone and ask what you want to do. You can ignore this prompt and proceed manually, or click it and tell Windows to copy the photo files. ■

TIP

Different Folder Names Some cameras and smartphones might use a name other than DCIM for the main folder. ■

If you have a digital camera, you can also copy photos to your computer from the camera's removable flash memory card. Just insert the memory card into your PC's memory card reader and proceed from there.

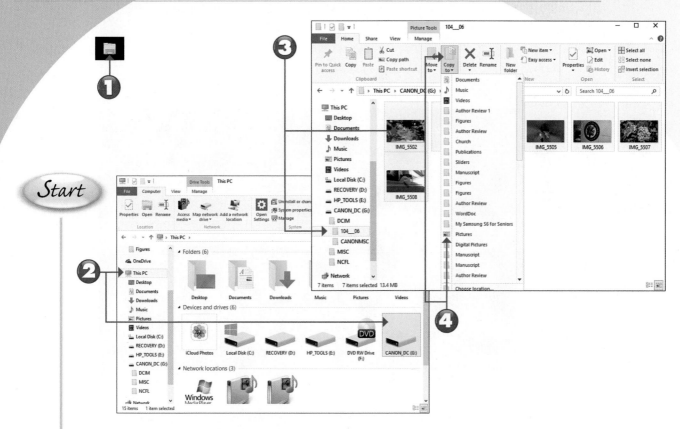

Start

1 Insert your camera's flash memory card into your PC's memory card reader and click **File Explorer** in the taskbar to open File Explorer.

2 Click **This PC** in the navigation pane and then double-click the drive for your memory card reader.

3 Navigate to and open the folder where the photos reside (usually labeled DCIM) and then hold down the **Ctrl** key and click each photo you want to transfer.

4 Select the **Home** ribbon and click **Copy To**, and then select **Pictures**.

End

NOTE

Copying Automatically Windows might recognize that your memory card contains digital photos and start to download those photos automatically—no manual interaction necessary. ■

CAUTION

Other Opening Apps Depending on what apps you have installed on your system, you might get multiple prompts to download photos when you connect your camera. If this happens, pick the program you prefer to work with and close the other dialog boxes. ■

VIEWING YOUR PHOTOS IN WINDOWS

When you want to view the photos stored on your PC, you can use the Photos app included with Windows 10. The Photos app organizes your photos, lets you view them one at a time or in a slideshow, and even lets you edit them. You open the Photos app from the Windows Start menu.

1 Open the Photos app from the Windows Start screen. By default, the Collection view is selected, and you see photos organized by date taken, wherever they happen to be stored.

2 To view photos by album, click **Albums**, and then click an album to view its contents.

3 Click through the folders and subfolders until you find a photo you want to view, and then click that photo to view it full-screen.

Continued

NOTE

OneDrive The Photos app automatically organizes and displays photos stored on your computer (and on all connected drives and devices), as well as those stored online in your OneDrive account. ■

4 To enlarge the picture, click the **+** button at the lower-right corner of the screen. To make a picture smaller, click the **-** button.

5 To move to the next picture in the collection, album, or folder, click the **right arrow** onscreen or press the **right arrow** key on your keyboard. To return to the previous picture, click the **left arrow** onscreen or press the **left arrow** key on your keyboard.

6 To view a slide show of the pictures in this folder, click the picture to display the menu bar at the top of the screen, and then click the **Slide Show** (play) button.

7 To delete the current picture, display the menu bar and click **Delete**.

End

TIP

Lock Screen Picture To use the current picture as the image on the Windows lock screen, open the photo, click to display the menu bar, click **Options** (three dots), and then click **Set as Lock Screen**. ■

TIP

Sharing Pictures To share the current picture with your friends, click to display the menu bar, click the **Share** icon, and then select to share via either Facebook or Mail. ■

EDITING YOUR PHOTOS WITH THE PHOTOS APP

Not all your pictures turn out perfectly. Maybe you need to crop a picture to highlight the important area. Maybe you need to brighten a dark picture, or darken a bright one. Or maybe you need to adjust the tint or color saturation. Fortunately, you're in luck—you can do all these basic touch-ups within the Windows Photos app.

Start

From within the Photos app, open the picture you want to edit, and then click it to display the options bar at the top of the screen.

The Photos app might try to automatically enhance the picture; if so, you see the Enhance tool selected. If you don't like the results, click **Enhance** to deselect this tool. (Click **Enhance** again to automatically enhance the original picture.)

To rotate the picture clockwise 90 degrees, click **Rotate**.

To further edit the picture, click **Edit**.

Continued

NOTE

Non-Destructive Editing Any changes you make are applied to a copy of the photo. The original unedited version of the photo is retained on your computer in case you ever want to revert to it. ■

5 Click **Basic Fixes** to access basic fixes.

6 Click **Crop** to crop the edges of the picture. When the crop screen appears, use your mouse to drag the corners of the white border until the picture appears as you like, and then click **Apply**.

7 Click **Red Eye** to remove the red-eye effect from the picture. The cursor changes to a blue circle. Move the circle to the eye(s) you want to fix, and then click the mouse button to remove red-eye.

8 Click **Retouch** to smooth out or remove blemishes from the photo. The cursor changes to a blue circle. Move the circle to the area you want to repair, and then click the mouse button to do so.

Continued

TIP

Aspect Ratio By default, Windows maintains the original aspect ratio when you crop a photo. To crop to a different aspect ratio, click the **Aspect Ratio** button and make a new selection. ■

NOTE

Red-Eye Red-eye is caused when a camera's flash causes the subject's eyes to appear a devilish red. Removing the red-eye effect involves changing the red color to black in the edited photo. ■

9 Click **Filters** to apply special photo filters to the picture. Click the desired filter on the right side of the window.

10 Click **Light** to edit the brightness and contrast of the photo.

11 Click the control you want to adjust—**Brightness**, **Contrast**, **Highlights**, or **Shadows**.

12 The selected control changes to a circular control. Click and drag the control clockwise to increase the effect, or counterclockwise to decrease the effect.

Continued

NOTE

Lighting Controls The Brightness control makes the picture lighter or darker. The Contrast control increases or decreases the difference between the photo's darkest and lightest areas. Use the Highlights control to bring out or hide detail in too-bright highlights; use the Shadows control to do the same in too-dark shadows. ■

13 Click **Color** to edit the tint and saturation of the photo.

14 Click the control you want to adjust—**Temperature**, **Tint**, **Saturation**, or **Color Boost**.

15 The selected control changes to a circular control. Click and drag the control clockwise to increase the effect, or counterclockwise to decrease the effect.

16 To apply vignette and selective focus effects to your picture, click **Effects** in the left sidebar and then click the effect you want to apply.

End

NOTE

Color Controls The Temperature control affects the color characteristics of lighting; you can adjust a photo so that it looks warmer (reddish) or cooler (bluish). The Tint control affects the shade of the color. The Saturation control affects the amount of color in the photo; completely desaturating a photo makes it black and white. And the Color Enhance control lets you click an area of the photo to increase or decrease color saturation. ■

Chapter 16

PROTECTING YOUR COMPUTER

"An ounce of prevention is worth a pound of cure" is a bit of a cliché, but it's also true—especially when it comes to your computer system. Spending a few minutes a week on preventive maintenance can save you from costly computer problems in the future.

To ease the task of protecting and maintaining your system, Windows includes several utilities to help you keep your computer running smoothly—and recover your data in case of some sort of malfunction.

WINDOWS ACTION CENTER

Notifications ————

Quick Actions ————

USING THE WINDOWS ACTION CENTER

The Windows 10 Action Center is the quick way to find out what's happening on your computer system and take immediate actions of various sorts. You can use the Action Center to turn on or off your PC's wireless functionality, access various configuration options, and switch to Tablet mode.

Start

1 Click the **Notifications** icon on the taskbar to display the Action Center.

2 The top section of the Action Center displays new system notifications, new email messages, and similar notices.

3 The bottom section displays Quick Actions; click a tile to initiate a given action.

4 Click **All Settings** to display the Settings window.

End

NOTE

Control Panel Another useful utility is the Windows Control Panel, which runs on the desktop and offers control of many system settings. To open the Control Panel, right-click the **Start** button and scroll until you can click **Control Panel**. ■

Computer viruses and spyware (collectively known as malicious software, or *malware*) install themselves on your computer, typically without your knowledge, and then either damage critical system files or surreptitiously send personal information to some devious third party. You can protect your system from viruses and spyware by using an anti-malware program, such as Windows Defender, which is built into Windows 10.

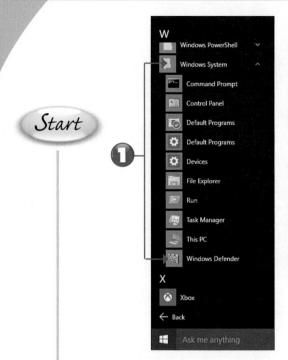

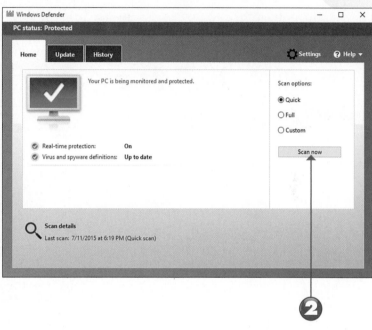

① Windows Defender runs in the background, monitoring your computer against malware threats. To open Windows Defender, click the **Start** button to open the Start menu, click **All Apps**, scroll to and click to open the **Windows System** folder, and then click **Windows Defender**.

② Defender automatically scans your system on its own schedule, but you can perform a manual scan at any time by clicking the **Scan Now** button.

End

TIP

Other Anti-malware Utilities Your computer manufacturer might substitute or supplement Windows Defender with other antivirus utilities, such as AVG AntiVirus (www.avg.com), Kaspersky Anti-Virus (www.kaspersky.com), McAfee AntiVirus Plus (www.mcafee.com), and Norton AntiVirus (www.symantec.com). Other anti-spyware utilities include Ad-Aware (www.lavasoftusa.com) and Spybot Search & Destroy (www.safer-networking.org). ■

CAUTION

How to Catch a Virus Computer viruses and spyware are most commonly transmitted via infected computer files. You can receive virus-infected files via email or instant messaging, by downloading files from unsecure websites, or by clicking links in Facebook or Twitter that link to malware-infested sites. ■

DELETING UNNECESSARY FILES

Even with today's humongous hard disks, you can still end up with too many useless files taking up too much hard disk space. Fortunately, Windows includes a utility that identifies and deletes unused files. The Disk Cleanup tool is what you should use when you need to free up extra hard disk space for more frequently used files.

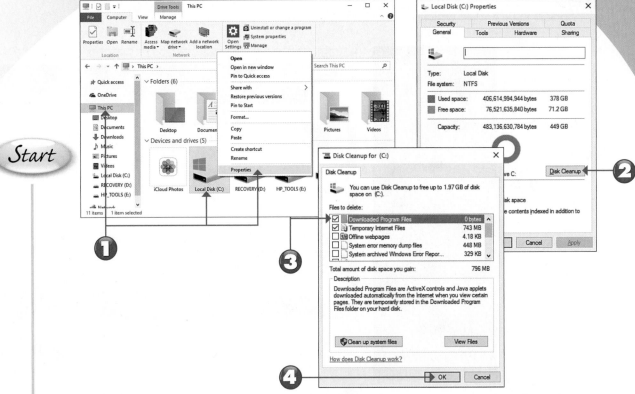

 Click File Explorer on the taskbar to open File Explorer, navigate to the **This PC** section, right-click the drive you want to clean up (usually the C: drive), and click **Properties**.

When the Properties dialog box opens, select the **General** tab (displayed by default), and click the **Disk Cleanup** button.

Disk Cleanup automatically analyzes the contents of your hard disk drive. When it's finished analyzing, it presents its results in the Disk Cleanup dialog box. Select which types of files you want to delete.

Click **OK** to begin deleting the selected files.

End

TIP

Which Files to Delete? You can safely choose to delete all these files *except* the setup log files and hibernation files, which are needed by the Windows operating system. ■

Another way to free up valuable hard disk space is to delete those programs you never use. This is accomplished from the Settings tool.

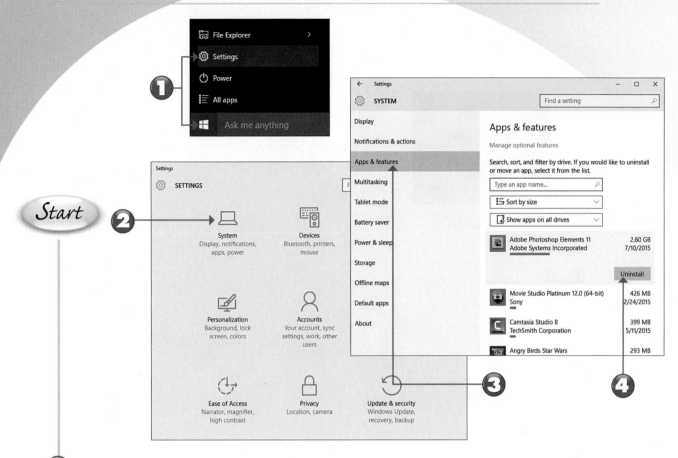

Start

1. Click the **Start** button and then click **Settings** to open the Settings window.

2. Click **System** to open the System page.

3. Click to select the **Apps & Features** tab.

4. Click the program you want to delete and then click **Uninstall**.

End

TIP

New PC Bloatware Most brand-new PCs come with unwanted programs and trial versions installed at the factory. Many users choose to delete these "bloat-ware" programs when they first run their PCs. ■

BACKING UP YOUR FILES

The data stored on your computer's hard disk is valuable, and perhaps irreplaceable. That's why you want to keep a backup copy of all these valuable files, either on an external hard disk connected to your PC or online to your OneDrive account.

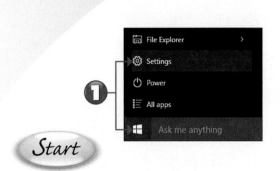

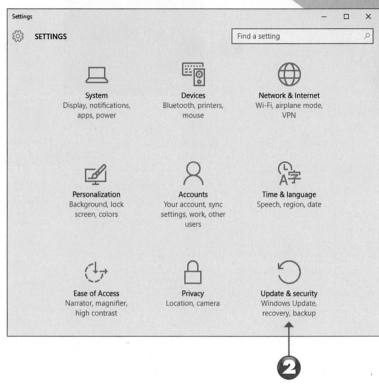

① Click the **Start** button and then click **Settings** to open the Settings window.

② Click **Update & Security** to open the Update & Security page.

Continued

TIP

External Hard Drives To create a local backup, purchase and install an external hard disk drive. These drives provide lots of storage space for a relatively low cost, and they connect to your PC via USB. ■

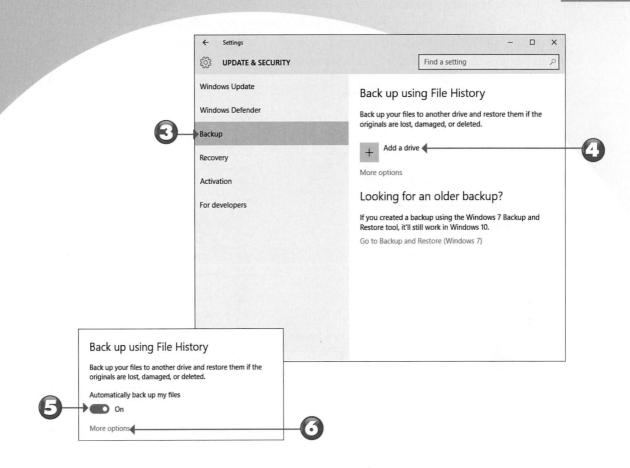

3 Click to select the **Backup** tab.

4 Click **+ Add a Drive** to select the drive or device where you want to store your backup copies.

5 File History is automatically activated. To turn off this automatic backup, click the **Automatically Back Up My Files** option to the Off position.

6 To select which folders are backed up, and how often, click **More Options**.

End

RESTORING YOUR COMPUTER AFTER A CRASH

If your computer system ever crashes or freezes, your best course of action is to run the System Restore utility. This utility can automatically restore your system to the state it was in before the crash occurred—and save you the trouble of reinstalling any damaged software programs. It's a great safety net for when things go wrong!

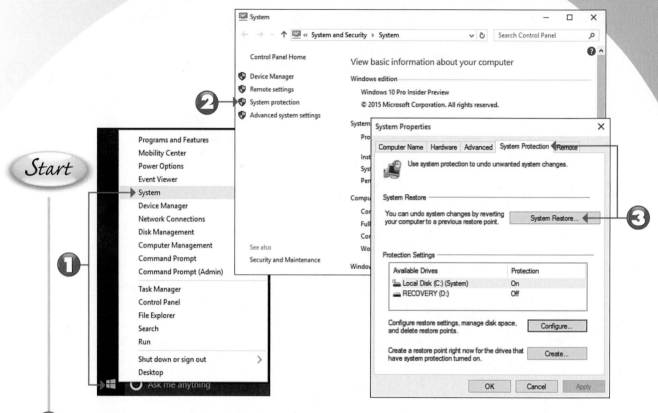

 Right-click the **Start** button and click **System** to display the System window.

 Click **System Protection** in the navigation pane to display the System Properties dialog box.

3 Make sure that the **System Protection** tab is selected, and then click the **System Restore** button.

Continued

TIP

Restoring Your System Be sure to close all programs before you use System Restore because Windows will need to be restarted when it's done. The full process might take half an hour or more. ■

CAUTION

System Files Only—No Documents System Restore will help you recover any damaged programs and system files, but it won't help you recover any documents or data files. This is why you need to use the File History utility to back up all your data on a regular basis—and restore that backed-up data in the case of an emergency. ■

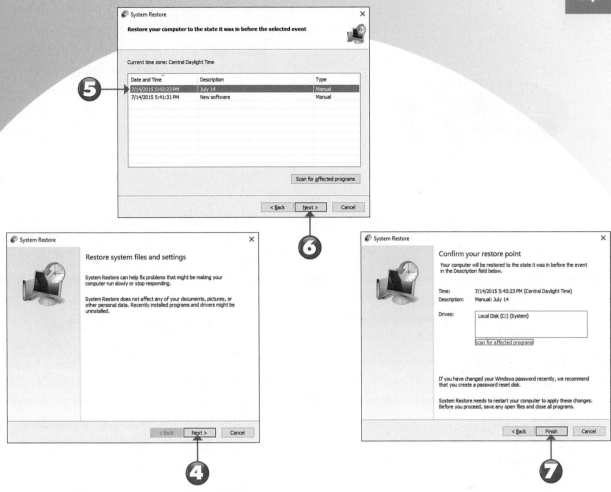

When the System Restore window appears, click **Next**.

Select a restore point from the list.

Click **Next**.

Click the **Finish** button to begin the restore process.

End

TIP

Refreshing System Files Windows 10 lets you "refresh" your system with the current versions of important system files, in case those files become damaged or deleted. Click **Update & Security** from the Settings window. On the next screen, select the **Recovery** tab, go to the Reset This PC section, and click **Get Started**. When prompted, click the **Keep My Files** option. ■

TIP

Resetting Your System In the event of a catastrophic system problem, you can reset your system to its factory-fresh condition by wiping clean the hard disk and reinstalling Windows from scratch. Click **Update & Security** from the Settings window. On the next screen, select the **Recovery** tab, go to the Reset This PC section, and click **Get Started**. When prompted, select the **Remove Everything** option. Note, however, that this option deletes all the programs and files on your computer—use it only in the case of an emergency. ■

Glossary

A

Action Center   The pop-up pane that appears when you click Notifications in the Windows taskbar; it displays system messages and quick links to key Windows functions.

add-in board   A device that plugs into a desktop computer's system unit and provides auxiliary functions. (Also called a *card*.)

address   The location of an Internet host. An email address might take the form johndoe@xyz.com; a web address might look like www.xyztech.com. See also *URL*.

all-in-one computer   A desktop computer in which the system unit, monitor, and speakers are housed in a single unit. Often the monitor of such a system has a touchscreen display.

app   See *application*.

application   A computer program designed for a specific task or use, such as word processing, accounting, or missile guidance.

attachment   A file, such as a Word document or graphics image, attached to an email message.

B

backup   A copy of important data files.

boot   The process of turning on your computer system.

B

broadband   A high-speed Internet connection; it's faster than the older dial-up connection.

browser   A program, such as Internet Explorer or Google Chrome, that translates the Hypertext Markup Language (HTML) of the Web into viewable web pages.

bug   An error in a software program or the hardware.

C

cable modem   A high-speed, broadband Internet connection via digital cable TV lines.

card   Also called an *add-in board*, this is a device that plugs into a desktop computer's system unit and provides auxiliary functions.

CD-ROM (compact disc read-only memory)   A CD that can be used to store computer data. A CD-ROM, similar to an audio CD, stores data in a form readable by a laser, resulting in a storage device of great capacity and quick accessibility.

computer   A programmable device that can store, retrieve, and process data.

Cortana   The virtual assistant built into Windows 10, designed to provide personalized information and search capabilities.

CPU (central processing unit)   The group of circuits that direct the entire computer system by (1) interpreting and executing program instruction and (2) coordinating the interaction of input, output, and storage devices.

cursor   The pointer that tracks with the movement of your mouse or arrow keys onscreen.

D

data   Information—on a computer, in digital format.

database   A program for arranging facts in the computer and retrieving them—the computer equivalent of a filing system.

desktop The graphical user interface within Windows on which running apps appear.

desktop computer A personal computer designed for use on a typical office desktop. A traditional desktop computer system consists of a system unit, monitor, keyboard, mouse, and speakers.

device A computer file that represents some object—physical or nonphysical—installed on your system.

disk A device that stores data in magnetic or optical format.

disk drive A mechanism for retrieving information stored on a magnetic disk. The drive rotates the disk at high speed and reads the data with a magnetic head similar to those used in tape recorders.

domain The identifying portion of an Internet address. In email addresses, the domain name follows the @ sign; in website addresses, the domain name follows the www.

download A way to transfer files, graphics, or other information from the Internet to your computer.

driver A support file that tells a program how to interact with a specific hardware device, such as a hard disk controller or video display card.

DSL (digital subscriber line) A high-speed Internet connection that uses the ultra-high-frequency portion of ordinary telephone lines, allowing users to send and receive voice and data on the same line at the same time.

DVD An optical disc, similar to a CD, that can hold a minimum of 4.7GB, enough for a full-length movie.

E

Edge The new web browser included with Windows 10.

email Electronic mail; a means of corresponding with other computer users over the Internet through digital messages.

encryption A method of encoding files so that only the recipient can read the information.

Ethernet A popular computer networking technology; Ethernet is used to network, or hook together, computers so that they can share information.

executable file A program you run on your computer system.

F

favorite A bookmarked site in a web browser.

file Any group of data treated as a single entity by the computer, such as a word processor document, a program, or a database.

File Explorer The utility used to navigate and display files and folders on your computer system. Previously known as Windows Explorer.

FiOS A type of broadband Internet service delivered over fiber-optic cable.

firewall Computer hardware or software with special security features to safeguard a computer connected to a network or to the Internet.

FireWire A high-speed bus used to connect digital devices, such as digital cameras and video cameras, to a computer system. Also known as *i.LINK* and *IEEE-1394*.

folder A way to group files on a disk; each folder can contain multiple files or other folders (called *subfolders*). Folders are sometimes called *directories*.

freeware Free software available over the Internet. This is in contrast with *shareware*, which is available freely but usually asks the user to send payment for using the software.

G

gigabyte (GB)   One billion bytes.

graphics   Pictures, photographs, and clip art.

H

hard disk   A sealed cartridge containing a magnetic storage disk(s) designed for long-term mass storage of computer data.

hardware   The physical equipment, as opposed to the programs and procedures, used in computing.

home page   The first or main page of a website.

homegroup   A small network of computers all running Windows.

hover   See *mouse over*.

hybrid computer   A portable computer that combines the functionality of a touchscreen tablet and a traditional notebook PC.

hyperlink   A connection between two tagged elements in a web page, or separate sites, that makes it possible to click from one to the other.

I-J

icon   A graphic symbol on the display screen that represents a file, a peripheral, or some other object or function.

Internet   The global network of networks that connects millions of computers and other devices around the world.

Internet service provider (ISP)   A company that provides end-user access to the Internet via its central computers and local access lines.

K-L

keyboard   The typewriter-like device used to type instructions to a personal computer.

kilobyte (KB)   A unit of measure for data storage or transmission equivalent to 1024 bytes; often rounded to 1000.

LAN (local-area network)   A system that enables users to connect PCs to one another or to minicomputers or mainframes.

laptop   A portable computer small enough to operate on one's lap. Also known as a *notebook* computer.

LCD (liquid crystal display)   A flat-screen display in which images are created by light transmitted through a layer of liquid crystals.

M-N

megabyte (MB)   One million bytes.

megahertz (MHz)   A measure of microprocessing speed; 1MHz equals one million electrical cycles per second.

memory   Temporary electronic storage for data and instructions, via electronic impulses on a chip.

microprocessor   A complete central processing unit assembled on a single silicon chip.

modem (modulator demodulator)   A device capable of converting a digital signal into an analog signal, typically used to connect to the Internet.

monitor   The display device on a computer, similar to a television screen.

motherboard   Typically the largest printed circuit board in a computer, housing the CPU chip and controlling circuitry.

mouse   A small handheld input device connected to a computer and featuring one or more button-style switches. When moved around on a flat surface, the mouse causes a symbol on the computer screen to make corresponding movements.

mouse over   The act of selecting an item by placing your cursor over an icon without clicking. Also known as *hovering*.

network   An interconnected group of computers.

notebook computer   A portable computer with all components (including keyboard, screen, and touchpad) contained in a single unit. Notebook PCs can typically be operated via either battery or wall power.

O-P

Office   Microsoft's suite of productivity applications—Word, Excel, PowerPoint, and more.

OneDrive   Microsoft's cloud-based storage service.

operating system   A sequence of programming codes that instructs a computer about its various parts and peripherals and how to operate them. Operating systems, such as Windows, deal only with the workings of the hardware and are separate from software programs.

path   The collection of folders and subfolders (listed in order of hierarchy) that hold a particular file.

peripheral   A device connected to the computer that provides communication or auxiliary functions.

phishing   The act of trying to "fish" for personal information via means of a deliberately deceptive email or website.

pixel   The individual picture elements that combine to create a video image.

port   An interface on a computer to which you can connect a device, either internally or externally.

printer   The piece of computer hardware that creates hard copy printouts of documents.

Q-R

RAM (random access memory)   A temporary storage space in which data can be held on a chip rather than being stored on disk or tape. The contents of RAM can be accessed or altered at any time during a session but will be lost when the computer is turned off.

resolution   The degree of clarity an image displays, typically expressed by the number of horizontal and vertical pixels or the number of dots per inch (dpi).

ribbon   A toolbarlike collection of action buttons, used in many Windows programs.

ROM (read-only memory)   A type of chip memory, the contents of which have been permanently recorded in a computer by the manufacturer and cannot be altered by the user.

root   The main directory or folder on a disk.

router   A piece of hardware or software that handles the connection between your home network and the Internet.

S

scanner   A device that converts paper documents or photos into a format that can be viewed on a computer and manipulated by the user.

screensaver   A display of moving designs on your computer screen when you haven't typed or moved the mouse for a while.

server   The central computer in a network, providing a service or data access to client computers on the network.

shareware   A software program distributed on the honor system; providers make their programs freely accessible over the Internet, with the understanding that those who use them will send payment to the provider after using them. See also *freeware*.

software   The programs and procedures, as opposed to the physical equipment, used in computing.

spam   Junk email. As a verb, it means to send thousands of copies of a junk email message.

spreadsheet   A program that performs mathematical operations on numbers arranged in large arrays; used mainly for accounting and other record keeping.

spyware   Software used to surreptitiously monitor computer use (that is, spy on other users).

Start menu   The pop-up menu, activated by clicking the Start button, that lists all installed programs on a computer. Microsoft removed the Start menu in Windows 8/8.1 but returned it—in an enhanced fashion—in Windows 10. Microsoft sometimes refers to the Windows 10 Start menu as the Start Experience.

system unit   The part of a desktop computer system that looks like a big beige or black box. The system unit typically contains the microprocessor, system memory, hard disk drive, floppy disk drives, and various cards.

T–U–V

tablet computer   A small, handheld computer with no keyboard or mouse, operated solely via its touchscreen display.

terabyte (TB)   One trillion bytes.

touchscreen display   A computer display that is touch sensitive and can be operated with a touch of the finger.

trackpad   The pointing device used on most notebook PCs, in lieu of an external mouse.

ultrabook   A type of small and thin notebook computer with no built-in CD/DVD drive and a smaller display.

Universal app   A type of application designed to run on a variety of Microsoft operating systems on different types of devices—PCs, tablets, and smartphones.

upgrade   To add a new or improved peripheral or part to your system hardware. Also to install a newer version of an existing piece of software.

upload   The act of copying a file from a personal computer to a website or an Internet server. The opposite of *download*.

URL (uniform resource locator)   The address that identifies a web page to a browser. Also known as a *web address*.

USB (universal serial bus)   The most common type of port for connecting peripherals to personal computers.

virus   A computer program segment or string of code that can attach itself to another program or file, reproduce itself, and spread from one computer to another. Viruses can destroy or change data and in other ways sabotage computer systems.

W–X–Y–Z

web page   An HTML file, containing text, graphics, and/or mini-applications, viewed with a web browser.

website   An organized, linked collection of web pages stored on an Internet server and read using a web browser. The opening page of a site is called a *home page*.

Wi-Fi   The radio frequency (RF)-based technology used for home and small-business wireless networks and for most public wireless Internet connections. It operates at 11Mbps (802.11b), 54Mbps (802.11g), 600Mbps (802.11n), or 1Gbps (with 802.11ac). Short for "wireless fidelity."

window   A portion of the screen display used to view simultaneously a different part of the file in use or a part of a different file than the one in use.

Windows   The generic name for all versions of Microsoft's graphical operating system.

Windows Explorer   See *File Explorer*.

Windows Store   Microsoft's online store that offers Universal apps for sale and download.

World Wide Web (WWW)   A vast network of information, particularly business, commercial, and government resources, that uses a hypertext system for quickly transmitting graphics, sound, and video over the Internet.

Zip file   A file that has been compressed for easier transmission.

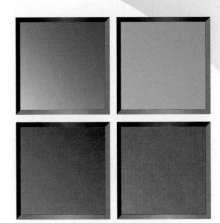

Index

F

G

H

N

O

Q-R

S

X-Y

Z

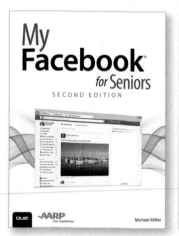
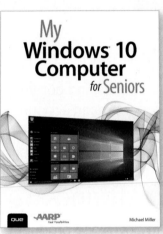

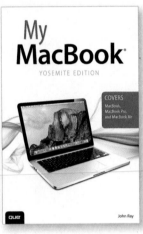
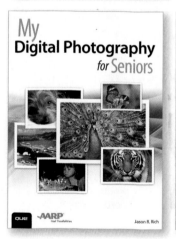